The Philosophy of
Set Theory

For Jim
in memory of theses
'not dead, but sleeping'

The Philosophy of
Set Theory

*An Introduction to
Cantor's Paradise*

MARY TILES

Basil Blackwell

Basil Blackwell Ltd
108 Cowley Road, Oxford, OX4 1JF, UK

Basil Blackwell Inc.
432 Park Avenue South, Suite 1503
New York, NY 10016, USA

British Library Cataloguing in Publication Data

Tiles, Mary
The philosophy of set theory: an
historical introduction to Cantor's
paradise.
1. Set theory
I. Title
511.3'22
ISBN 0-631-15285-7

Library of Congress Cataloging in Publication Data

Tiles, Mary.
The philosophy of set theory: an historical introduction to Cantor's paradise/Mary Tiles.
p.   cm.
Bibliography: p.
Includes index.
ISBN 0-631-15285-7: $55.00 (U.S.)
1. Mathematics–Philosophy.   2. Set theory.   I. Title.
QA8.4.T54   1989   88-26065
511.3'22–dc 19   CIP

Typeset in 11 on 13 pt Times
by Joshua Associates Limited, Oxford
Printed in Great Britain by
Bookcraft Ltd., Bath, Avon

Like the enigma of time for Augustine, the enigma of the continuum arises because language misleads us into applying to it a picture that doesn't fit. Set theory preserves the inappropriate picture of something discontinuous, but makes statements about it that contradict the picture, under the impression that it is breaking with prejudices; whereas what should really have been done is to point out that the picture just doesn't fit, that it certainly can't be stretched without being torn, and that instead of it one can use a new picture in certain respects similar to the old one.

Wittgenstein, 1974, p. 471

Contents

Preface

This book is the result of a number of attempts to teach under-graduate classes and seminars in the philosophy of mathematics, sometimes to a mixture of mathematicians and philosophers, sometimes just to interested philosophers. I have found it difficult to recommend reading that was neither too mathematically technical for the philosophers nor too philosophically technical for the mathematicians. I confess to failing wholly to resolve that problem, although that was the aim. The hope is nevertheless that both philosophers with only a very basic grounding in mathematics and mathematicians who have taken only an introductory course in philosophy may find something of interest by differentially skipping over those parts which are either too technical or too familiar. I would like in particular to persuade philosophers that the philosophy of mathematics is not an isolated speciality but is inseparably intertwined with what are standardly regarded as mainstream philosophical issues.

Thanks are due to Dr Jim Brown of Toronto University for encouraging the project and for test running the draft version, and to his students for their comments and corrections; also to Hal Switkay for pointing out some mathematical errors. I would like to thank also Swarthmore students David Ravinsky and Russell Marcus for their suggestions and for letting me try out material on them. Above all thanks to Jim Tiles for all those innumerable forms of support without which the book would never have been completed.

Introduction: Invention or Discovery?

Did Cantor discover the rich and strange world of transfinite sets (which Hilbert was to call Cantor's Paradise) or did he (with a little help from his friends) create it? Are set theorists now discovering more about the universe to which Cantor showed them the way, or are they continuing the creative process? Perhaps they are wandering in a wonderland which is no more understandable and no more substantial than that in which Alice, in Lewis Carroll's *Alice in Wonderland*, found herself.

One way of approaching these questions is to think about them in relation to a question to which Cantor devoted much time in his later years: 'How many points are there in a line?' Cantor thought he knew the answer. 'There are $2^{\aleph_0}$ and $2^{\aleph_0} = \aleph_1$' ($\aleph$, aleph, is the first letter of the Hebrew alphabet.) Here $\aleph_0$ is the first infinite cardinal number. It is the number of the set of natural numbers, $\{0, 1, 2, 3, \ldots\}$, irrespective of the order in which they are counted. $2^{\aleph_0}$ is to be understood by analogy with $2^2, 2^3, 2^4$, etc. and $\aleph_1$ is the next largest infinite cardinal number after $\aleph_0$. $2^{\aleph_0} = \aleph_1$, which has come to be known as Cantor's continuum hypothesis, thus says that the number of points on a line is the second infinite cardinal number: there are none in between $\aleph_0$ and $2^{\aleph_0}$. Notice that we would not reach this conclusion by generalizing from 2^n. For example $8 (= 2^3)$ is not the next number after 3, and in general 2^n is not the next number after n. But infinite numbers are strange things and we should not expect them to behave in all respects like finite numbers.

Cantor's problem was that, although convinced of the correctness of his hypothesis, he never succeeded in proving it. Moreover, subsequent work in set theory has not resolved the question. We may know a lot more about the problem, but we also know that

Cantor's continuum hypothesis cannot be proved from the standardly accepted axioms of set theory. What then, should be our attitude towards his hypothesis and toward possible answers to the question concerning the number of points in a line? This all depends on the answer given to the questions with which we started.

Suppose Cantor did *discover* a new realm, a realm which has now been more extensively explored and systematically mapped since the axiomatization of set theory. Then his hypothesis, being a hypothesis about things in this realm, must be either correct or incorrect even though our present axioms do not characterize this realm sufficiently precisely for a proof to be given which would enable us to determine which is the case. Only by discovering new axioms, new fundamental truths about the set theoretic universe, will it be possible to give proofs which would settle the matter. But how are these new axioms to be discovered? How do we gain access to the uncharted areas of this realm? These would be questions which could not be avoided.

On the other hand, suppose that Cantor *created* the realm of infinite sets and infinite numbers in the way that Lewis Carroll created Wonderland. Although creations of this sort can be discussed, analysed and schematized by others (we might try constructing a map of Wonderland), they are nevertheless such that there will be some questions about them which simply have no answers because the creator did not supply enough information to provide an answer and there is no other source of information. The question 'What was the diameter of the top of the Madhatter's hat?' has no answer, for *Alice in Wonderland* neither includes this directly in its description nor provides details of other dimensions from which an answer could be deduced. Given that this is so, we could consistently add to the story by filling in this detail (in the way that illustrators have filled in the price of the hat – Tenniel gives it a price tag of $10/6d$, whereas that in Rackham's illustration is $8/4d$) and we might do so in different ways, so continuing the creative process. Of course we might also argue that dimensions in Wonderland are peculiarly problematic; given Alice's tendency to grow and shrink we would have to specify the frame of reference carefully. It seems that the continuum hypothesis is a similarly unanswerable question about infinite numbers. There are several different ways in which we can add to the basic set theoretic axioms, so giving rise to different extended set theories, each of which would be seen as a filling out of the original. There are some constraints on what number is assigned to the continuum, but there is also a very considerable degree of freedom. If mathematics is a free creative activity, constrained only by demands of consistency, absence of contradiction, then all of these alternative extended set theories have an equal claim to mathematical legitimacy.

It might, however, be argued that the continuum hypothesis was not originally asked as a question about a self-contained realm of infinite sets created by Cantor. It was asked in the course of attempting to answer questions about the points forming a line. When this is taken into account it might be more plausible to suppose that Cantor *invented* his infinite numbers and that he did so for the purpose of solving problems concerning the characterization of continuous spaces and functions defined on them, problems which arose out of the concern to provide infinitesimal calculus with a rigorous foundation. Inventions frequently have to be refined and improved in the course of being put to work. Since both the continuum hypothesis and its negation are consistent with the basic axioms of set theory, any decision on it should be based on what proves to be most helpful in resolving the problems which the theory of infinite numbers was designed to solve. To this end all alternatives and their implications for areas of mathematics outside set theory should be explored. It may be found that one of these alternatives is to be preferred, or that different alternatives are useful for different purposes. (Here we should note that the line between invention and discovery cannot easily be made sharp. In the natural sciences inventions, such as the light bulb, are often also described as discoveries. This is because discovery includes discovery of ways of doing things as well as of the existence of previously unknown things, and discovering new ways of doing things frequently involves inventing new instruments.)

Finally, the whole situation might be interpreted as evidence that talk of infinite numbers is not really to be taken seriously. There are those who would insist that talk of the infinite always was, and still is, nonsense. Cantor's continuum hypothesis is neither true nor false because it makes no sense. Moreover, there is no meaningful statement of the form 'the number of points on a line is . . .' so there is no such statement which can be either true or false.

Discussion of the continuum hypothesis cannot therefore be separated from the wider, essentially philosophical questions about the status of set theory in general and of the theory of transfinite numbers in particular. But how is there to be any adjudication between the philosophic positions just sketched? On what basis could one hope to find an answer to the questions with which this Introduction began? They are not the sort of questions on which a direct assault will yield much progress. The indirect approach to be adopted here is as follows: We shall start by seeking to understand the most radical of the options suggested above – that of thinking that all talk of infinite numbers and of the number of points on a line is nonsense. This was the orthodox position from the time of Aristotle until well into the seventeenth century. Then we shall consider how a person starting from such a view might be convinced that there is some sense which can be given to this talk by examining what sense it was given and how. We can then address the question of which, if any, of the less radical positions this account of sense, or arguments based on it, would justify. The aim here is not to attack or undercut the idea of the actual infinite, but to explore it by finding out how and why it became mathematically important. This is an indirect way of shedding light on the sense in which the domain of transfinite sets and numbers might be thought to constitute a reality.

The adoption of this indirect approach is motivated by the feeling that simply to opt for a strong realist position, asserting that all claims about the universe of sets, including the continuum hypothesis, are determinately true or false, whether or not we have any means of knowing which, does not help to determine the nature of that reality or to elucidate the means by which we may acquire knowledge concerning it. It does not increase our understanding of what mathematicians are about when they are doing set theory. But equally, to adopt a strong finitist position and to deny the legitimacy of all this talk of the transfinite will still leave us in a position of having to give an explanation of the mathematical activity which constitutes classical set theory and of all the other mathematics which makes appeal to its results. In either case, to argue at a purely philosophical level for some general realism or some general anti-realism will leave all of the substantial work of understanding existing mathematical activity yet to be done. The aim of what follows is, however, the limited one of giving pointers to the sort of framework within which such an understanding might be sought, by focusing on the role and status of the actual infinite.

The intention is to focus on philosophic issues rather than on technical details, which have therefore been kept to a minimum, with the inevitable result that there can be no pretensions to formal rigour. Readers wishing to check up on the formal details are advised to follow up the references included in the text and the suggestions for further reading. Chapter 8, on the independence results, is the most technical and can be omitted by those who are prepared to take the independence of the generalized continuum hypothesis and of the axiom of choice from the remaining Zermelo–Fraenkel set theory on trust. On the other hand, those who have already taken a course on set theory might wish to omit chapter 6, on the axiomatization of set theory.

1

The Finite Universe

Infinite, or transfinite numbers and transfinite set theory are relative newcomers on the mathematical scene. Cantor's most important papers on the theory of transfinite numbers, the culmination of work begun in 1870, were published in 1895 and 1897 (Cantor, 1915). Thus, if one were to proclaim them to be inventions, figments of mathematical imagination, one would not be casting aside centuries of tradition. Indeed, the weight of tradition is firmly opposed to giving credence to talk of any such things. The infinite only gained acceptance and a degree of mathematical respectability because traditional ways of thinking were being cast aside.

Also the revolution has not been complete. We are still more likely to be suspicious of talk of infinite numbers and infinite sets than of talk of the familiar whole numbers and fractions that are used in counting and in the simple computations which are an essential part of many practical activities and all commercial transactions. These misgivings about any theory of transfinite sets or transfinite numbers are reflected by those philosophers who would accept the label 'finitist'. There are *prima facie* two possible types of finitism:

1 Finitism

Strict Finitism   The strict finitist does not recognize any mathematical use of the infinitistic notions or of infinistic methods (summing an infinite series for example). The strict finitist might also want to distinguish between 'small' and 'large' finite numbers, arguing that there is a (finite) upper limit on the numbers with which we can deal intelligibly, although there may be much debate about how any such limit can be set or determined.

One route to such a position is explored in Wright (1980) which elaborates on themes suggested by Wittgenstein (1967). The basic claim here is that we can only know of the existence of those numbers which we could actually write down in some notation and 'take in' all at once, or survey. Similarly, it is suggested, we can only be convinced by a proof which we can survey. It is quite possible for a purported proof to be too long and too complex for us to take in (for example a computer generated 'proof' running to a hundred pages). If such a sequence cannot be taken in, it cannot be a proof, for it cannot perform the function of a proof, which is to convince us that its conclusion is true (given agreed assumptions which form the premises of the proof) by showing us why it must be true. On this basis it may be supposed that there is an upper bound on the natural numbers (it just is not true that every number has a successor even though it would be impossible to specify one which does not, since if we can specify n we can specify $n + 1$). This bound will be set by our cognitive powers coupled with the efficiency of our system of notation. Models of this situation are afforded by computing systems whose upper limits are imposed by the memory size together with the structure of the software.

It is clear, however, that any precise statement of a strict finitist position will be a delicate matter. The distinction between 'small' (surveyable) finite numbers and 'large' (unsurveyable) finite numbers has much in common with the distinction between men who are bald and men who are not; the distinction is real even though the loss of a single hair cannot effect the transition from not being bald to being bald. Similarly the strict finitist will need to say that the distinction between surveyable and unsurveyable numbers is real even though the addition of 1 is not sufficient to effect the transition to unsurveyability. (Further possible motivations for pursuing this position will emerge in chapter 2, but these are essentially linked to arguments which appear to close off the option of classical finitism.)

(Classical) Finitism   The (classical) finitist is quite happy about the mathematical status of the familiar natural numbers, however large, but refuses to accept the need for infinite numbers or sets, and

indeed does not regard talk of such things as coherent. However, he does not dismiss all notions of infinity or all mathematical treatments of infinite series. His insistence is that such things are only *potentially*, not actually, infinite; any actual segment of such a series is always finite, but always incomplete. It is in this incompletability that its *potential* infinity consists. The mathematician, when dealing with these always incomplete, potentially infinite series, must thus use methods which differ from those used when dealing with completed or completable finite series.

Finitists of both types argue not only that we do not need infinite numbers or a theory of infinite sets, but also that experience affords us no basis on which to give sense to talk of them. We do not need infinite numbers or infinite sets because all applications of mathematics are to finite systems, finite quantities and finite numbers of entities (see Hilbert, 1925). Since any application involving measurement can only ever be approximate, in view of the fact that every measuring instrument, however good, has a built in margin of error, we only ever need a finite number of decimal places when assigning a numerical value to a physical magnitude. Moreover, it may be argued, our experience is that of finite beings and takes the form of finite sequences of impressions of entities which are also finite. There can thus be no empirical meaning given to talk of the actual infinite. Following this line of argument, some empiricists have been led to conclude that there is no sense to be given to such talk. These claims clearly rest on (a) an assumption about the finitude of the universe within which mathematics is applied, (b) an assumption that mathematics is only applied to this universe via processes of measurement, and (c) an assumption that meaning is to be equated with empirical meaning.

This route to making out the finitist case makes it rest heavily on empiricist doctrines about meaning, doctrines which have been seriously challenged, as a result of the failures of logical positivism, by work which follows in the wake of Quine's 'Two Dogmas of Empiricism' (1953). As so formulated it is therefore unlikely to be taken seriously by philosophers advocating realism as the general position to be adopted in the theory of meaning. If we take it that a necessary (but not sufficient) condition of realism with respect to statements of a given kind is that it is held that all statements of this kind are determinately true or false independent of our ability to know which is the case (cf. Dummett, 1963), it will be clear that the realist will not be impressed by arguments against the infinite which appeal to restrictions on our cognitive capacities imposed by either the finiteness of our intellects or the finite character of all experience. In general he will be prepared to allow our ability to conceptualize and entertain possibilities to outrun our capacity (even to know in principle) which, if any, of these possibilities are ever actualized. But what case can the realist make which might persuade the finitist (an anti-realist about the infinite), motivated by empiricism, of the error of his ways?

There are two challenges to which the finitist position, as outlined above, is open. It may be conceded that in our measuring of features of the physical world finite numbers and finite strings of decimals will always serve, but if the finitist thinks of his measurements and observations as measurements and observations of features of a physical world, then he is making two assumptions which require him to think both beyond the finite and beyond the bounds of experience conceived as a sequence of impressions. First he presumes that what he encounters are items extended in space and existing for some, possibly very short, period of time. In doing so he makes at least implicit use of the notion of continuity, or of continuous extension, for space and time are presumed to be continuous. And the notion of continuity brings with it that of infinite divisibility. Secondly he presumes that the things he observes and measures are all parts of a single physical world and can all be located in a single spatio-temporal framework. This all-embracing character of space, time and the universe suggests not only that they must be thought of, even though they are in no sense observable, but also that they must be thought to be infinite since neither space nor time can coherently be thought to have a boundary. So, it would seem, talk of space and time already takes us beyond what can be given content by reference to immediate experience and already threatens to introduce the infinite.

We shall see that, in order to respond to these challenges, the finitist (unless he is prepared to reject the continuity of space and time or to deny the possibility of giving any empirically significant theoretical account of it), must abandon strict finitism and must admit that some sense can be given to talk of the infinite, but without allowing that this legitimates talk of infinite numbers. He will insist that the only possible sense of 'infinity' which can be grounded in experience, more liberally construed as including a reflective awareness of rules and principles, is that of the potentially infinite. It will be argued below that there is a contradiction involved in thinking that a number can be assigned to a potential infinity or in thinking that it forms the sort of determinate collection that a set is required to be. Thus what the finitist has to do is to show that the challenges arising out of the continuity and the unity of space and time can be handled by invoking only the notion of potential infinity. He has to argue that he is not, in virtue of his presuppositions about space and time, committed to the supposition that there are actually, in the physical universe, undetectable infinitely small and infinitely large quantities, or any actually infinite sets of points. Moreover, the route taken by the argument to be considered will reveal grounds other than those tied to forms of empiricism, verificationism or global anti-realism for advocating a finitist position. It will thus suggest that realism about the infinite is no automatic consequence of a generally realist stance elsewhere.

2 Continuity and Infinity

Whereas the question 'How many points are there in a line?' could at least have been asked prior to Cantor, his answer, 'There are $2^{\aleph_0}$ and $2^{\aleph_0} = \aleph_1$' could not even have been understood or recognized as an answer. It was Cantor who introduced and defined the symbols '$\aleph_0$' and '$\aleph_1$' as symbols for infinite (cardinal) numbers. We cannot begin to understand his answer without knowing how these symbols are defined, and to understand their definitions it is necessary to know something about the theoretical background which legitimates them as definitions. For we first have to be convinced that there are, or at least might be, such things as infinite numbers for which we can introduce names. Thus it is only within the framework provided by transfinite set theory that it becomes possible to contemplate giving a *numerical* answer to our question; this framework provides the form of an answer, if not an actual answer. To this extent Cantor's work *gives sense* to a question which previously lacked any precise sense.

Prior to Cantor the natural answer would have been 'Infinitely many', and if the question 'And how many is that?' were further pressed, it would have been taken as showing a lack of understanding of what is meant in this context by 'infinitely many'. This is not to say that finitism was the only possible position prior to Cantor's work; it is just that the non-finitist would not have been able to give an answer couched in terms of transfinite numbers, or any other numbers.

One could take Berkeley as a representative finitist and Leibniz as a representative non-finitist. Berkeley insisted that space can only ever (actually) be infinitely divided and was highly critical of Newton's infinitistic methods (Berkeley, 1734). Leibniz was well aware of the distinction between potential and actual infinites and believed that matter is actually infinitely divided (Leibniz, 1702). He was prepared to admit the existence of infinitesimal magnitudes and himself developed an infinitesimal calculus at much the same time as Newton. But it is clear that Leibniz could have attached no more sense than Berkeley to a question about the number of points in a line. Here there is not only a question about whether 'Infinitely many' is a legitimate answer to a 'How many?' question but also about whether the points on a line form a totality of which one can sensibly ask 'How many are there?'

One might suggest, as Wittgenstein (1967, p. 59) does, that such a question has as much or as little sense as 'How many angels can dance on a needlepoint?' The problem is that one is simply unclear as to how to determine the totality to be numbered. It is not clear that even one angel can dance on a needlepoint, whereas there are points in lines. But a line is a single, continuous whole. How can it be made up of points? If one point is to be continuously adjoined to another, it must be no distance from it and therefore must in fact coincide with it. Alternatively, given any two distinct points there will always be some distance between them, and so also more points between them. This suffices to show that no finite number can be assigned to the points in a line because one will always be able to find more points in between those counted.

This is merely another way of saying that a continuous line is infinitely, i.e. indefinitely, divisible, for points are points of division. The points in a line mark the boundaries of parts (actual or potential) of the line. If a line is potentially infinitely divisible (but only

ever actually finitely divided), there must be a corresponding
potential infinity of points of division. If the line is actually infinitely
divided, there must be an actual infinity of points of division. But
neither from the point of view of the finitist with respect to actual
division, nor from that of the non-finitist do the points in a line form
the sort of totality to which one could, in principle, assign a number.
The answer 'Infinitely many' is thus both an assertion that there are
more than any given finite number and a refusal of the demand for a
numerical answer.

That this is the right response appeared to have been con-
clusively demonstrated by Aristotle in his treatment of Zeno's
paradoxes (Aristotle, *Physics*, Bk. II). Zeno's paradoxes are
important because they seem to show that it is simply impossible to
think of a continuum, such as space or time, as made up out of
indivisible atoms, whether these be points or minimally extended
regions of space and/or time. There is a contradiction between the
existence of movement and an atomistic conception of space and
time. But we know that movement exists. This is not something we
can deny, therefore it is the possibility of an atomistic view of
continua such as space and time which must be rejected. This is not
just a human impossibility but a logical impossibility, and therefore
one which any philosopher, whether of realist or anti-realist
persuasions, must accept as indicative of a limitation on what is
conceptually possible.

These paradoxes have, naturally, been discussed at length (see,
for example, Salmon, 1967). However, the point of introducing
them here is not that of presenting a new resolution but of showing
how one of their oldest resolutions, that indicated by Aristotle, gave
support to a finitist position – the position which rejects all talk of
infinite sets, infinite numbers, etc., as unnecessary nonsense;
something which could never form part of mathematics conceived
as a rational scientific discipline. In later chapters we shall see how
the Aristotelian view had to be rejected in order that the case for the
infinite could be made out.

3 Zeno's Paradoxes

The strategy of Zeno's paradoxes may be reconstructed in the
following way (modified from Owen, 1957).

Thesis I   Neither space nor time are pluralities. For if they are pluralities it must be possible to specify the units (atomic parts) of which they are composed. But

Thesis II   Any attempt to treat space or time as *composed* of atomic parts leads to absurd conclusions. For suppose they are composed of atomic parts, then a space or a time must *either* be divisible without limit *or* there must exist limits of division.

A   Suppose they are divisible without limit, then
   1 a runner cannot complete a racecourse, and
   2 Achilles cannot catch the tortoise.
B   Suppose there are limits of division, then *either* these have size (magnitude) *or* they do not.
   (a) Suppose they have size, then
   3 the paradox of the stadium.
   (b) Suppose they have no size, then
   4 the arrow paradox.

Thus the alternatives lead to absurd conclusions, and neither space nor time are pluralities.

   1   Space and time have been supposed to be composed of parts and to be divisible without limit. Consider then the plight of a runner trying to run the full course of a race (figure 1.1). Before he can get to the end he must get to the half-way point. But then before he can get to the end he must get to the half-way point of the remaining section, and so on without end. There is thus no end to the sequence of stages which he must complete before reaching the finishing post; there will always be a bit more of the course to be run and he can therefore never reach the finishing post.

   2   Again, consider a race between Achilles and a tortoise (figure 1.2). Suppose Achilles gives the tortoise a 10 m start and that he runs twice as fast as the tortoise. Say Achilles runs at 2 m per second and the tortoise at 1 m per second. Then in 10 s Achilles

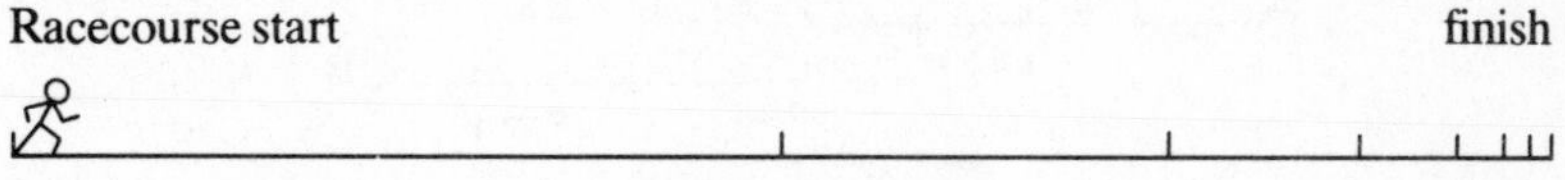

Racecourse start   finish

Figure 1.1

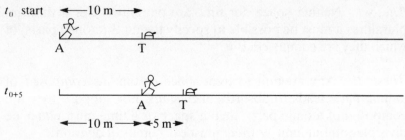

t_0 start

t_{0+5}

Figure 1.2

will have covered 20 m and the tortoise will have covered 10 m, i.e. Achilles will have caught up with the tortoise in 10 s. But Zeno argues that if space is divisible without limit and is composed of its parts, then Achilles has first to reach the tortoise's original position (which he will do in 5 seconds) and by then the tortoise will have gone ahead 5 m. Achilles will take another 2.5 seconds to cover that 5 m and in that time the Tortoise will have moved ahead 2.5 m. Thus, whenever Achilles reaches the tortoise's position, no matter how little time it takes him, the tortoise will have moved some distance ahead, and Achilles must always reach the tortoise's old position before catching him up. The tortoise will always be ahead of Achilles and Achilles can never catch him.

3   Suppose then that space and time are not divisible without limit but that there are limits of division and that these have size – i.e. there are indivisible, extended space units and time units. Consider the situation shown in figure 1.3. Suppose A and B are moving with equal speeds, $v = 1$, i.e. both at the rate of one space unit (atom) per time unit (atom), but in opposite directions. Then after one time unit we will have the situation shown in figure 1.4. A and B have each moved one unit relative to the stadium but two units relative to each other. But if B has moved two units relative to

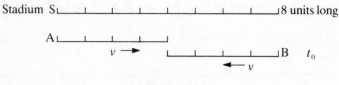

Figure 1.3

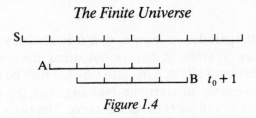

Figure 1.4

A there must have been a time when it had moved just one unit relative to A, i.e. our time unit must be divisible after all and therefore cannot be a limit of division. Moreover, when B has moved just one unit relative to A it must have moved half a unit relative to the stadium. Thus, after all, our spatial unit is not indivisible either.

4 We now suppose that the limits of division have no size. Consider the arrow loosed from the archer's bow (figure 1.5). At a moment t (a limit of temporal division having no size) the arrow must occupy a space which is precisely equal to its own volume. Everything, when it occupies a space equal to its own size, is at rest. Therefore the arrow is at rest at t. But the whole time of its flight is composed of moments, thus at each moment during its flight it is at rest. So the arrow is at rest throughout its flight and never moves.

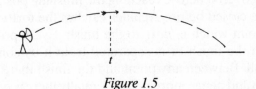

Figure 1.5

The conclusion which Aristotle draws from Zeno's paradoxes is that neither space nor time nor any other continuous magnitude may be supposed to be composed, made up out of, or considered as an aggregate of parts. Although all such magnitudes have parts, and indeed it is characteristic of them to be divisible without limit, they are not built up out of those parts. Rather, the parts exist as parts only as a result of division of the whole of which they are parts; continuous wholes are wholes which are given before their parts.

Aristotelians could thus deal with the first two paradoxes in the following way: It is correct to see space and time as divisible without limit. What the paradoxes show is that it is absurd to think that where there are no limits of division, no atoms, the divided whole can be built up out of all the parts into which it *may* be divided, for (a) there is nowhere to start the building, and (b) there is no determinate totality of parts of such a whole. The parts are created by division. We could have considered successive division by three, or by any other number instead of by two.

Another way of saying that a continuum is indefinitely divisible is to say that it can be divided *anywhere*, i.e. there are no 'natural' divisions, divisions which correspond to its constitution out of parts. A continuum forms a whole because it is homogeneous and not in itself differentiated into parts; any such differentiation is imposed. Zeno can deduce his absurd conclusions only by making it look as if in order to run the length of a racecourse a runner must *successively* build his journey out of parts which are such that there can be no last. Similarly for Achilles narrowing the gap between himself and the tortoise.

From this point of view, however, it would also follow that the course, not only cannot be completed, but also that the runner cannot even start to run it. For Zeno's arguments rely on the fact that divisibility without limit entails that there is no *last* part of the course to be covered before reaching the finishing post, there is no *last* gap to be closed before Achilles catches the tortoise, because there is no point which is *next* to the finish. Every point is either identical with the finish or is separated from it by some distance however small. Between any point and the finish there will always be a finite gap and hence more points. Equally there is no point next to the start (figure 1.6).

If A is to start, then he must be at rest at one instant and moving at the next. But if A *is* moving at *t* then he must have already moved some distance, for if he is still at the start he is not yet moving. Thus he must have been moving and there must be a time before *t* at which he was also moving, and a time before that, and so on. How, then, can motion ever begin? The whole point is that there is no

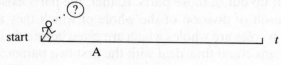

start

A

Figure 1.6

next instant and no *next* point to reach. We cannot count off instants of time or points on a line *in order*, and are only tempted to think that we can by a false 'atomistic vision'.

The runner can traverse the racecourse, and in doing so he will traverse all its potential parts and will thereby pass all the points at which the course might be divided, but he traverses the parts *by* or *in* traversing the whole, he does *not* traverse the whole by traversing the parts. For almost all of these parts (those not marked by marker posts, etc.) are not actual; they do not actually exist but are a misplaced concretization (one which is not coherently conceptualizable or imaginable) of the potentiality for division which expresses the type of homogeneity which is called continuity.

This response to paradoxes 1 and 2 dictates at least part of the response to paradoxes 3 and 4. Since divisibility without limit is accepted, the supposition behind 3 and 4 that there are limits of division is rejected. Paradox 3 can then be straightforwardly taken to show the impossibility of regarding continua such as space and time as being made up out of extended atoms – indivisible chunks which have size. For in being continuously extended they will necessarily be potentially divisible (this is part of the conceptual content of extension) and the potential is required to be actualizable if it is to be possible for there to be bodies moving in opposite directions (or even, as Russell argued (Russell, 1903, § 322), for there to be bodies moving with different velocities).

Aristotle's treatment of 4, the arrow paradox, is the most controversial in that it has been held responsible for blocking the path to a science of dynamics for nearly 2000 years. And it is certainly from the success of the new dynamics of Galileo, Descartes and Newton that the pressure to drop the Aristotelian finitist perspective comes. In dealing with the arrow paradox the Aristotelian will first deny that points are *parts* of space and that instants are *parts* of time; they are not limits of division and cannot be reached by division. No extended whole, no continuous magnitude can be made up out of what has no magnitude, for adding together two things of no magnitude cannot increase their size. Points, lines, planes, etc. are boundaries or limits between things which do not themselves have an existence independent of the things of which they form the boundaries or limits. Time and space are not composed of such limits, but boundaries may be introduced to mark off one time from another, one region of space from another and may really exist in the sense that the boundary is marked and is detectable, for example by there being some genuine contrast between the things on either side of the boundary.

There is a boundary between black and white and it has a determinate, more or less rectangular shape (figure 1.7). But if the black were removed leaving only the boundary we would simply have a uniformly white surface; the boundary does not survive the elimination of the colour discontinuity (lack of homogeneity) between regions which created, or brought with it, the spatial division. Similarly no sense could be made of the suggestion that the black rectangle is in fact such that all points on the line joining x and y (a proper mathematical line with no thickness) are green. Points and lines cannot be coloured, only surfaces or regions.

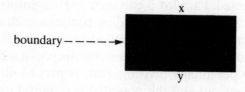

Figure 1.7

This illustrates part of the sense in which points are said not to be *parts* of spatial regions (and instants not to be parts of temporal intervals). If R is a uniformly coloured black surface, all parts of R are also uniformly black. It is this form of reasoning, transferred to the case of time and motion, which forms the basis of Aristotle's response to the arrow paradox. For, Aristotle argues (*Physics* VI, 8), from the fact that at any instant *t* during its flight the arrow does not move, it does not follow that the arrow is at rest at *t*. For, he says, motion and rest are terms which apply only in relation to periods of time. Since there can be no motion in an instant (all movement takes time) there can be no rest in an instant either. To qualify for being at rest, as opposed to being in motion, an object must occupy the same place for a period of time. It is just as absurd to suppose that the arrow moved throughout a time interval *T* except for an instant *t* at which it was at rest, or that it was at rest throughout T except for an instant t^* at which it had velocity v, as to have a black surface divided by a green line.

Aristotle then faults Zeno's reasoning on two counts: (a) For the arrow to be moving throughout T it must indeed be moving throughout all parts of T. But since instants dividing T are not parts of T, this does not require that the arrow should be moving for each instant. Therefore it can be agreed that there is no motion in an instant without having to conclude that there is no motion at all. (b) Again, because T is not composed of instants, motion during T is not the sum of motions in the instants dividing T. Therefore it does not follow from the fact that at each instant there is no motion, that there is no motion at all. Thus Aristotle says:

> It is true that at any now it [a moving object] is always over against something; but it is not at rest; for at a now it is not possible for anything to be either in motion or at rest. So while it is true to say that that which is in motion is at a now not in motion and is opposite some particular thing, it cannot in a period of time be at rest over and against anything; for that would involve the conclusion that that which is in locomotion is at rest. (1984, *Physics*, VI, 8, 239a33–239b4)

> Zeno's reasoning, however, is fallacious, when he says that if everything when it occupies an equal space is at rest, and if that which is in locomotion is always occupying such a place at any moment, the flying arrow is therefore motionless. This is false, for time is not composed of indivisible moments any more than any other magnitude is composed of indivisibles. (*Physics*, VI, 9, 239b5–10)

So far, so good. However, Aristotle's conclusions hold only on condition that infinite velocities are ruled out. For the claim that there can be no motion in an instant holds only if it is also assumed that all motion takes some (finite time). Instantaneous motion might be accomplished if things could ever move at infinite velocities. Infinite velocities and instantaneous motions are not perhaps wholly contradictory, but to suppose that any material object could behave in this way would require many other well entrenched assumptions about these objects to be abandoned. Suppose an object had the trajectory shown in figure 1.8. Where is M at t? It seems that either M does not have a position at t or that it is in two positions at t, which violates the assumption that material objects occupy exactly one place at one time and occupy some place at all times. Contemporary physics would at least support Aristotle to the extent that it holds that nothing can travel faster than light, which has a finite velocity of 186 000 miles per second.

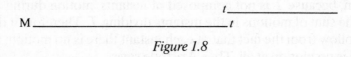

Figure 1.8

The controversial part of Aristotle's reasoning here is that in refusing to allow the characterization of instants in terms of motion or rest he effectively blocks the formation of the concept of instantaneous velocity (or of a state of motion) which is so crucial to all post-seventeenth-century physics. From our standpoint we could say that Aristotle fails to· distinguish between motion *in* an instant, which we agree to be impossible given that infinite velocities are impossible, with motion *at* an instant, which gives the state of motion of a moving body (one which is moving for a period of time containing that instant). Whether Aristotle's treatment of Zeno was the sole or even principle obstacle to the formation of the concept of instantaneous velocity is not a matter that need be pursued here. For the resolution of the difficulty raised by Zeno need rest solely on reasoning concerning motion *in* an instant.

With these responses to Zeno, Aristotle presents a powerful case for supposing that continuous magnitudes such as space and time not only need not, but cannot, be regarded as composed of atomic parts, whether these are finite or infinite in number. They must, on the other hand, be recognized to be divisible without a limit. In this restricted sense continuity may be said to bring a notion of infinity in its wake – the notion of potential infinity which is associated with the idea of a never-ending process, or an operation which is indefinitely repeatable, such as that of adding one to a whole number. But in such cases there is no infinite collection of actual parts of space or time and no actual completed infinite totality of numbers.

If there is no need for the introduction of infinite collections or infinite numbers in relation to continuous magnitudes, the finitist has a fairly strong case. He will, like Aristotle (and like Hilbert, 1925) rest his case on the fact that in our dealings with the real, physical world, the fundamental reality by reference to which all discourse ultimately gets its meaning, we never have any need for, or even any use for, the notion of infinity. We could not directly observe anything infinitely large or infinitely small but all our observation does take place in space and time and is of what is spatially and/or temporally extended (objects, their properties, changes and relations to one another, and events – flashes, bangs and other happenings). If, in order to think coherently about space, time and observable, spatio-temporally extended objects and their changes, including changes of position, it is necessary to suppose that space and time are composed of infinitely small parts, or that they are infinite sets of points, then the finitist will find it hard to make out his case. For the infinite, if not required directly by observation, would none the less be required to give coherent expression to the spatio-temporal framework within which it is assumed that all observation takes place. This is why Aristotle's response to Zeno is important.

His refusal of any actual infinity associated with continuity is made possible by insisting:

1 Continuous wholes are divisible without limit.
2 Continuous wholes are given prior to their parts and are not therefore composed of (are not totalities of) the parts into which they are potentially divisible.
3 Indefinite divisibility is a reflection of a distinctive kind of homogeneous unity possessed by continuous wholes, unity and homogeneity not possessed by anything composed out of 'atomic' parts.
4 Continuity must therefore be taken as a primitive, irreducible notion.
5 Only what can be formed by division is a part of a continuous magnitude. Points are therefore not parts of a line nor instants parts of time; they are boundaries, divisions which have only a potential existence unless actualized as discontinuities of some kind between the regions which they divide.

4 The Universe and the Absolutely Infinite

Even if the continuity of space and time can be acknowledged by the finitist and can be understood in terms of a potential infinity of divisions, there remains that aspect of our thought about space, time and the physical world which seems to require the employment of the notion of an actually existing infinite totality. For the universe itself, or if not this then space and time at least, must surely be infinite. Admittedly none of these are items encountered in experience, but in order to be able to conceptualize our experience as experience of a physical world we have to relate all those experiences by treating them as experiences of things which all form part of a single world, all stand in determinate spatio-temporal relations to each other. Can that all-embracing whole, the physical world, or the universe, be anything short of infinite? If not, cosmology and astronomy would be examples of physical sciences in which the notion of the infinite finds application, is given physical significance and must be given significance if our ordinary thought about experience as experience of a single world is to be possible.

Aristotle would have sought to counter such an argument by admitting the need to talk of the universe whilst insisting that the physical universe is spatially finite, although temporally infinite. However, the sense in which time is allowed to be infinite is that at any time there always was a preceding time and will always be a succeeding one. In other words there is, for Aristotle, no actually existent (or subsistent) entity, time; there is only a never-ending sequence of physical processes which constitutes the passage of time. Time, then, is only potentially infinite. (There is a problem about the status of the past, for Aristotle, for although one might allow that the future is potentially infinite, since it has not yet come to be, the past has already come to be and so, if it had no beginning, would seem to be actually, rather than potentially, infinite.)

However, as the Stoics pointed out, the idea that the universe is spatially finite is problematic. Simplicius, a sixth-century commentator on Aristotle puts the point in the following way:

(1) The Stoics want there to be a void outside the world and prove it through the following assumption. (2) Let someone

stand at the edge of the fixed sphere and stretch out his hand upwards. (3) If he does stretch it out, they take it that something exists outside the world into which he has stretched it, and if he cannot stretch it out, there will be something outside which prevents him from doing so. (4) And if he should next stand at the limit of *this* and stretch out his hand, a similar question will arise (5) For something which is also outside that point will also have been indicated. (SVF 2.535) (Long and Sedley, 1987, p. 49)

This is illustrated in figure 1.9.

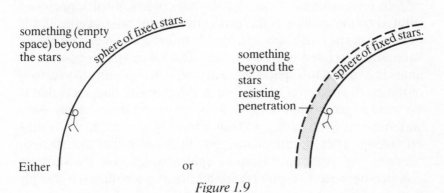

Figure 1.9

(This passage was much discussed in the middle ages and for a full and interesting account of the history of discussions of the possibility or otherwise of the extracosmic void see Grant (1969).)

The basic thought is that even if the universe (the material bit) is spatially finite, does this not simply mean that there must be empty space beyond the edge of the material universe, for how can space itself be only finite in extent? A finite spatial region has a spatial boundary, a limit. But spatial boundaries are always boundaries between one region of space and another, so space itself can surely have no bound, no limit. This thought can be backed up by making explicit another property which is seen as being almost as fundamental to continuous magnitudes as indefinite divisibility. Given any two magnitudes x, y (figure 1.10) such that either $x < y$ or $y < x$ (i.e. two comparable magnitudes, magnitudes of the same kind, two lengths, two areas, etc.) then there are finite integers n

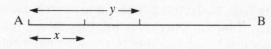

Figure 1.10

and *m* such that $nx > y$ and $my > x$. (Sometimes known as the Archimedean axiom although it is already mentioned in Aristotle in *Physics*, (III, 206b3–13).)

However, some caution is required here. The question of having bounds or not must be distinguished from that of being finite or infinite in magnitude. Consider the two-dimensional space constituted by the surface of the earth (simplifying we can consider it to be the surface of a sphere). This is finite; its area can be calculated and is related to its curvature. But there are no bounds or limits internal to this space. We will never, in exploring the surface of the earth encounter an edge of it. What would happen is that if we were to go in what we took to be a straight line, due east say, and were to keep going without changing direction, we would eventually, after a finite time, get back to where we started. 'Straight-line' paths in this space (paths which give the shortest distance between two points) are all finite in length, even though they need have no end points because they are all either parts of circles or circles (closed curves).

We now see that the Archimedean property has two readings: a spatial one according to which it is always possible to get beyond B starting from A by taking *x*-sized steps, however small *x* may be. This is true on the surface of the earth (provided 'steps' is not taken too literally!). The other reading concerns magnitudes. This would say that, given magnitudes *x*, *y* such that $y > x$, there is always a magnitude $mx > y$, which has the effect of saying that there is no greatest magnitude of the kind to which *x*, *y* belong. And this is not true for straight line lengths on the surface of the earth. If *y* were the length of the equator (A and B here coincide) then there would be no longer straight-line path on the surface of the earth. There can of course be a longer straight-line *journey*: the traveller can go round more than once.

The Aristotelian could then maintain his view of the spatially finite universe by saying that our space is the three-dimensional analogue of the two-dimensional surface of the sphere – the three-dimensional 'surface' of a four-dimensional 'sphere'. (This would take one very close to the Einsteinian view of the space–time universe as finite but unbounded, where space–time itself is 'curved', i.e. the geometry of space–time, like the geometry of the surface of the sphere, is not Euclidean.) Where the finitist has problems is if physical space, or space–time, is presumed to be Euclidean, for then it cannot be finite without also being bounded. The universe of classical Newtonian physics with its flat Euclidean space would have to be infinite. But with the displacement of a cosmology based on classical physics by one based on relativity theory, it would appear that the finitist could stand his ground.

However, if the finitist claim is that *there is no sense to* talk of infinite numbers, infinitely large magnitudes or infinite sets, it is not enough for him to argue from the *fact* that the universe is, or must be, finite on the basis of currently accepted physical theory. This is the way in which Aristotle (*Physics*, III) and Einstein (1920, chs. XXXI, XXXII) argue – from physics to finitude. But, as suggested above, not every physical theory would have this consequence; even if the universe is in fact finite, the possibility of classical physics would suggest that it might have been infinite and that it is at least coherent to suppose this. For did people not have to understand the classical Newtonian view of the universe in order to be able to see how and why it should be rejected in favour of Einstein's?

Here again we may learn from Aristotle's discussion, if not from the details of his cosmology. For Aristotle provides some of the conceptual tools which the finitist needs if he is to see how best to present his case. Moreover, these tools will in fact prove to be very necessary even for the Cantorian set theorist to present a consistent position.

The universe, the totality of what there is, if it is to be treated as an object, a whole, is none the less a whole, a totality of a very special kind, for it represents a kind of absolute maximum. There just isn't anything which could contain more or which could be bigger. And, as Aristotle points out, there is an incompatibility between the notion of a potential infinity and that of a totality which forms a complete whole, a unit. Indeed there is here the source of another notion of infinity – the absolutely infinite, that than which nothing can be greater. The universe is that of which every other (created) being is a limited part. The absolutely infinite is that by reference to which every other being is recognized to be limited and to fall short of the maximum. There can be no limit to the absolutely infinite for then it would not be maximal; it would be possible for something to surpass that limit and hence be greater. Yet the absolutely infinite is complete in that there is nothing which is not contained within it. (It is in this sense, for example, that the God of Descartes or Spinoza is said to be infinite. Descartes' God has all perfections and all to an unlimited extent; nothing could be more perfect in any respect. Spinoza's God-or-Nature has all possible attributes and is not limited in respect of any of them.)

Aristotle, although evidently aware of this sense of infinite, the infinity of an infinite *being* (substance – whether God, the Universe or Nature) does not think it a coherent notion. And it is possible to see where the problem lies, for there is a certain tension between completeness and unlimitedness – lacking limits or boundaries, between thinking of something as a unit, a whole thing, and thinking of it as infinite. This tension becomes an incompatibility if, as Aristotle thinks, the only notion of infinite, the only way in which sense can be given to being unlimited, is that of being potentially infinite. For the whole idea of a potentially infinite series is that it is never complete and can never be completed. Being unlimited is here given an essentially negative reading as necessary *lack* of completeness, or *lack* of completability. The concept of potential infinity is essentially linked to the idea of a process of construction, of generation, or simply coming to be. Thus Aristotle says:

> For generally the infinite has this mode of existence; one thing is always being taken after another, and each thing that is taken is always finite, but always different. (Again, 'being' is spoken of in several different ways, so that we must not regard the infinite as a 'this', such as a man or a horse, but must suppose it to exist in the sense in which we speak of the day or the games as existing – things whose being has not come to them like that of a substance, but consists in a process of coming to be or passing away, finite yet always different. (*Physics*, III, 206a26–34)

And

> The infinite turns out to be the contrary of what it is said to be: it is not what has nothing outside it that is infinite, but what always has something outside it. (206b35–207a1)

> Thus something is infinite if, taking it quantity by quantity, we can always take something outside. On the other hand, what has nothing outside is complete and whole. For then we define the whole – that from which nothing is wanting . . . Nothing is complete which has no end and the end is a limit. (207a7–14)

If Aristotle is right here, if the only viable sense of 'infinite' is that of the potentially infinite, then the universe must either be finite or not a completed whole, not a unity.

Let us go back to the model for a possible spatially finite universe in two dimensions – the surface of a sphere. This is finite precisely in the sense of not being potentially infinite; there is a finite maximum length to any 'straight' line path in this space (travelling always the shortest distance between two points, i.e. along a great circle). Travelling in such a straight line one cannot, taking constant sized steps keep covering new ground for ever; one step is taken after another and the ground covered is always finite but it is *not* always different. This space then is not infinitely large and, importantly, this is a conclusion which can be reached on the basis of the intrinsic character of the space; it does not require an external viewpoint from which boundaries are detected or set (from a third dimension). Its size is linked to its curvature and this in turn to the function which gives the distance between any two points. (See Nerlich (1976) for a further discussion of the characterization of space 'from the inside'.) This is important for, of course, the universe is, *ex hypothesi*, that to which there is no outside. If anything is to be said about it, it must be sayable from the inside, as it were.

Moreover, from this it follows that the universe, whether finite or infinite, cannot have the sort of completeness or wholeness which belongs to things within it, for these are such that their principle of completeness serves to mark them off from other things, *from* the rest of the universe, and thus enable a boundary to be drawn. The universe cannot be marked off from anything, cannot be limited by anything, if it contains all that there is. Even if the physical, spatio-temporal universe is not all that there is, it still cannot be limited by or marked off from anything spatio-temporal (since it contains all that is spatio-temporal); whatever else there is must be wholly different in kind from the things contained in the physical universe, something with which the physical universe can have no common boundary, something which is not spatio-temporal. This holds whether the universe is finite or infinite. If it is allowed to be a completed whole, a unit, it is a whole of a very special kind and it is this specialness which is captured by the positive sense of unlimited invoked by those who would insist on the legitimacy of the notion of an absolute or an actual infinity; it is that which is unbounded, not limited by anything external, within which all limits or boundaries and all finite (in the sense of limited) beings exist.

Potential infinity is opposed to the sort of completeness required of the universe, just because the universe consists of what is, of what is *actual*, whilst the potentially infinite is never fully actual but always in the process of becoming. It is an inescapably temporal notion. It is therefore perhaps not surprising that discussions of the allowable or comprehensible notions of infinity are intimately bound up with views on the nature of time, its 'reality' or otherwise, and with the opposition between metaphysics of 'being' and that of 'becoming'. To insist on the potentially infinite as the only legitimate infinite requires, in addition, an insistence on the reality of time and the importance of the notion of becoming. So that the finitist might well be committed to defending this kind of metaphysical position.

If the universe were to be allowed to be potentially infinite spatially, this would only be possible if space and time were not treated as independent. Space can be potentially infinite only if it is necessary to think of spatial distance, spatial separation in terms of tracing out paths in space; to think space potentially infinite can only mean that travelling in a straight line one will never come back to the same place but will forever be covering fresh ground. But if space is independent of time (space is the order of what coexists at a single time) it cannot be potentially infinite without also being actually infinite. It can only be possible continually to cover new ground because that ground is there to be covered. Time should not enter into the conceptual characterization of space as finite or infinite if all parts of space have to coexist at a time in order to be added to yield a spatial magnitude, whether finite or infinite. So if there is to be an infinite space, independent of time, it must be actually infinite space.

If the actual infinite makes no sense, then space must either be finite, or inseparable from time (as is the case in relativity theory where, although we may make local separations, no global separation is possible) i.e. the finitist would seem to have to deny the conceptual coherence of Newtonian space and time. The only further alternative position open to him is that adopted by Pascal (a finitist in a (more-or-less) Newtonian universe) who argued that a potential infinite always presupposes an actual infinite, and thus that since the created world is potentially infinite we thereby have proof of the existence of an actual infinite. However, our finite intellects cannot comprehend the actual infinite and are led into contradictions by attempting to do so. Thus although we can know that it exists, we also thereby know that there is a limit to what we can understand and hope to comprehend by the use of reason. This domain of necessary ignorance is the domain of faith (Pascal, 1954, pp. 589–91, 1105–07).

This argument of Pascal's is far from trivial, for it raises a very serious question about the viability of any finitist position which admits even the notion of potential infinity (i.e. any finitism which is not strict finitism). Cantor himself uses the same argument (quoted in Hallett, 1984, p. 25):

> There is no doubt that we cannot do without *variable* quantities in the sense of the potential infinite: and from this can be demonstrated the necessity of the actual-infinite. In order for there to be a variable quantity in some mathematical study, the 'domain' of its variability must strictly speaking be known beforehand through a definition. However, this domain cannot itself be something variable, since otherwise each fixed support for the study would collapse. Thus this 'domain' is a definite, actually infinite set of values.
>
> Thus each potential infinite, if it is rigorously applicable mathematically, presupposes an actual infinite.

If it is the case that any potential infinity presupposes an actual infinity there is evidently a problem. It would of course follow that time too would have to be actually, not merely potentially, infinite and thus in some sense wholly actual even though not simultaneously present. It is from this point of view that the reality of time as associated which change and becoming is questionable, and time as associated with change gives way to time as a fourth, quasi-spatial dimension.

What has to be recognized is that the line between potential and actual infinite is not easy to hold, even though, as we have seen, there are two distinct senses of infinite (unlimited) which ground the two notions. For when the geometer defines a circle as the locus of a point moving in a plane equidistant from a fixed point, he does not suppose that this motion literally takes place in time. He may imagine it as taking place in time, but in order to grasp that the path traced is a circle he also has to have a non-temporal conception of it as a complete path. Imagined generating slips into imagined completed generation; temporal becoming slips into atemporal being. For how else is the mathematician able to say anything about a potentially infinite domain, such as that of the natural numbers, or the points on a line. How can he theorize and generalize about the possibilities without supposing them to be, in some sense, actual and forming a completed totality. In other words, the finitist has to face two related questions: How is it possible to theorize about possibilities? and: What exactly is presupposed in the making of generalizations? How can a statement about all natural numbers be true if there is no completed totality of numbers in virtue of which it is true? How is it possible to generalize over a potentially infinite totality when it never is a totality?

It was all very well for Aristotle to say that by denying the existence of the actually infinite he was not depriving mathematicians of anything that they really needed because they only ever supposed that a line could be extended as far as they wished and so were only ever concerned with finite lines (*Physics*, III, 207b28–35). But this seems to ignore the fact that the mathematician is not only concerned with particular lines taken one at a time, but is concerned to prove theorems which hold for all possible lines of a given kind, all possible points on a line, etc. Whether the universe is actually finite or infinite may not make much difference at the practical level where we are dealing with individual objects, regions and their measurements, for there is always a margin of error such that we could never detect the difference between lines which, if indefinitely produced, would eventually meet and those which would not. But it does make a difference at the theoretical level when we come to develop the mathematical theory behind our measurement practices. The finitist case, if it is to be made out, therefore has to be linked to an account of generalization and of the meaning of universal statements. This is where we must turn to logic, and the theory of classes which was destined to play an important role in the early debates concerning the foundations of set theory.

2

Classes and Aristotelian Logic

The viability of classical finitism has been seen to rest on the possibility of providing a counter to Pascal's claim that every potential infinite presupposes an actual infinite. This is because the classical finitist wishes to be able to engage in traditional mathematical theorizing about the natural numbers, about points, lines, triangles, and so on. To theorize about these requires making assertions about all natural numbers, all points on a given line, etc. So the classical finitist needs to be able to give an account of the meaning of such generalizations which does not depend on the existence of the natural numbers as a completed (actually infinite) totality, or of the points on the line as a determinate, actually infinite set. He wishes to generalize over domains which he regards as necessarily incomplete.

The disagreement between Pascal and the classical finitist is closely related to a much wider, long-standing dispute between nominalists and essentialists, and before that between the Stoics and the Aristotelians, over the interpretation of formal logic, the meanings of general terms and hence over the account to be given of the meaning of statements of the form 'all As are Bs'. It was against the background of this dispute that the theory of classes received its earliest expositions. The notion of 'class' which arose from this tradition of logical and metaphysical debate forms part of the background of Cantor's own theory of classes.

Cantor's theory is now frequently called a 'naive theory of classes' (or a 'naive set theory'), for, with hindsight, it can be seen that it was an insufficiently critical employment of a notion of class, backed as it was by a long tradition which was responsible for many of the unclarities in Cantor's theory, and in particular for those which quickly yielded paradoxes. It was as a result of the challenge to provide a coherent, paradox-free account of numbers along Cantorian lines, based on classes, that the traditional Aristotelian logic came to be superceded by that of Frege (in the forms popularized by Russell, Hilbert, and others).

However, to understand how our question about the relation between the potentially and the actually infinite relates to, and depends on, issues about logic and the theory of classes, it is necessary to go back the Scholastic–Aristotelian framework.

1 Aristotelian Logic

The core of Aristotelian logic is the theory of syllogism. An example of the simplest form of syllogism would be:

All S are M.	All scholars are melancholic.
All M are P.	All melancholics are philosophers.
Therefore	Therefore
All S are P.	All scholars are philosophers.

A syllogism is a valid argument consisting of two premises and a conclusion, which between them contain just three terms, each of which occurs twice. The concern in syllogistic reasoning is to establish a relation between two terms (S and P) via their mutual relation to a third (or middle) term (M). Each sentence occurring in a syllogism thus asserts that a given relation holds between two terms, and there are four possible relations traditionally recognized:

All S are P.	No S are P.	and their negations
Some S are not P.	Some S are P.	

Thus another example of a syllogism would be:

All S are M.	All scholars are melancholic.
No P are M.	No philosophers are melancholic.
Therefore	Therefore
Some S are not P.	Some scholars are not philosophers.

For present purposes, it is not necessary to go into the full theory of syllogisms. It is sufficient to note that Aristotle (in his *Prior Analytics*) considered all the possible syllogistic forms and classified them into those which were valid (yielded syllogisms) and those which were not. It is part of the purpose of a theory of syllogisms, as of any philosophically motivated logical theory, to provide a means of discriminating between valid and invalid forms of argument by reference to (1) an account of the meanings of the logical constants involved, or (2) simple rules, which are either (a) accepted as being obviously correct, or (b) taken to be definitive of the logical constants. Aristotle adopts version (a) of (2), but much of the discussion of medieval and scholastic logicians was concerned with (1) and it is out of these discussions that the theory of classes as the extensions of terms eventually emerged.

On this account it is presumed that to each term there corresponds a class (the extension of the term) – the class of all things to which the terms correctly applies – and terms are treated as denoting, or standing for their extensions. This means that relations between terms are construed as grounded in relations between the classes they denote. The interpretation of the four possible relations between terms is then:

All *S* are P.	The class of *S*s is included in the class of *P*s.
No *S* are *P*.	The class of *S*s is excluded from the class of *P*s.
Some *S* are *P*.	The class of *S*s overlaps the class of *P*s.
Some *S* are not *P*.	The class of *S*s overlaps the class of things which are not *P*.

Euler, and later Venn, introduced diagrams to illustrate these relationships and to provide a diagrammatic means of testing the validity of syllogisms. Euler used circles to represent classes and gave the representations, shown in figure 2.1, of what is asserted in each of the above four cases.

The problem with these representations is that, with the exception of that for 'no *S* are *P*', the representation is not unique; there are several situations compatible with the truth of the statement and thus no one representation could be said to capture its exact

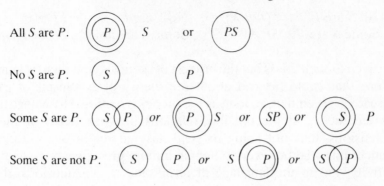

Figure 2.1

content. Venn diagrams, however, achieve a unique diagrammatic representation for each type of statement (figure 2.2).

The Venn diagrams achieve a unique representation for each type of statement by moving away from thinking of them as asserting relations between terms. Instead they represent the three possible classes definable by conjoining *S* with either *P* or its

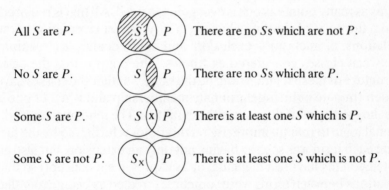

Figure 2.2

negation, *not-P*, and indicate whether they are said to be empty or non-empty. (Shading is used to indicate that a class is empty and 'x' to indicate that it is non-empty.) In this, Venn follows the analysis which Leibniz (1966, pp. 81–4) gave of the content of the four kinds of statement when developing his theory of syllogisms based on an algebra of terms. His algebraic representation was:

All S are P. $S(not\text{-}P) = 0$ No S are P. $SP = 0$
Some S are P. $SP \neq 0$ Some S are not P. $S(not\text{-}P) \neq 0$

Even though this is but the briefest of sketches, and there is much more that could be said about the theory of syllogisms, it can already be seen that, so long as the theory is confined to syllogisms as the theory of the possible relations between terms or their extensions, it is not going to yield any answer to our question concerning whether every potential infinity presupposes an actual infinity. Put in the language of terms and their extensions, this becomes the question 'Can a term, such as "natural number", have a potentially infinite extension, or are all classes, as the extensions of terms, actual, complete totalities?'

It can be seen from the fact that Venn diagrams provide a decision procedure for discriminating between valid and invalid syllogistic forms that this question does not need to be faced when dealing merely with the validity of syllogistic inferences. These diagrams represent classes (the extensions of terms) as continuous wholes (areas), leaving the representation of members of the classes quite indeterminate. They exploit the fact that a continuous whole contains as many points as one is ever going to need. All that is required, from the point of view of syllogistic theory, is a representation of the relations of inclusion, exclusion and their denials as relations between classes considered as wholes. The point is that the basic structure of these relations will be the same whether the classes have a determinate or an indeterminate membership and whether one is dealing with possibilities or with actualities. The tendency for traditional logic to remain imprecise on this point is furthered by the fact that each term was seen as having not only an extension, but also an intension (as having a meaning, or as expressing a concept) so that relations between terms were sometimes treated as being grounded in relations between their intensions. The traditional discriminations between valid and invalid syllogisms can be grounded in either way. This is made clear by Leibniz, whose algebra of terms can be given either an intensional or an extensional interpretation. We thus get the parallel readings set out in table 2.1.

Because there is this possibility of systematic double reading (extensional/intensional, existential/essential) the study of logic traditionally concerned itself with both intensional (or formal) relations between terms, and with extensional (or material) relations between terms, not without, however, frequent debates on their relative priority. For this reason the notion of class which grew out of this logical tradition is vague both as regards questions of determinate or indeterminate membership and on the exact relation between extension and intension. The Aristotelian logical core is neutral on such issues (and we shall see why in the next section). Thus it is not within logic as such that any clarification is to

Table 2.1

All A are B	$A(not\text{-}B) = 0$	The extension of $A(not\text{-}B)$ is empty
	(*extensional*)	The extension of A is included in the extension of B
	(*existential*)	Every (actual) A is (in fact) B
		$A(not\text{-}B)$ is a contradictory concept
	(*intensional*)	The intension of A includes the intension of B
	(*essential*)	Any (possible) A is (necessarily) B
No A are B	$AB = 0$	The extension of AB is empty
	(*extensional*)	The extension of A is included in that of *not-B*
	(*existential*)	No (actual) A is (in fact) B
		AB is a contradictory concept
	(*intensional*)	The intension of A includes that of *not-B*
	(*essential*)	No A could (possibly) be B
Some A are B	$AB \neq 0$	The extension of AB is not empty
	(*extensional*)	The extension of A is not included in the extension of *not-B*
	(*existential*)	There is an A which is B
		AB is not a contradictory concept
	(*intensional*)	The intension of A does not include the intension of *not-B*
	(*essential*)	It is possible for an A to be B

Table 2.1 (cont.)

Some A are not B	$A(\textit{not-}B) \neq 0$	The extension of $A(\textit{not-}B)$ is not empty
	(*extensional*)	The extension of A is not included in the extension of B
	(*existential*)	There is an A which is not B
		$A(\textit{not-}B)$ is not a contradictory concept
	(*intensional*)	The intension of A does not include the intension of B
	(*essential*)	It is possible for an A not to be B

Note that in moving from extensional to intensional readings the inclusion relations are simply reversed.

be sought. Rather, the presence of this tradition and its tendency to treat logic as an independent and rather elementary study is the source of a very vague notion of class. Here, because it is presumed that every term has both an extension and an intension, it is possible to slip from thinking of a class as a mere collection of objects to thinking of it as essentially dependent on a concept, which, by giving necessary and sufficient conditions of membership in the class, determines its limits and unites its members. This can be done without anyone noticing because it makes no difference to the syllogisms judged to be valid.

Moreover, with the usual spatial diagrams, classes are being thought of in terms of spatial regions which are assumed to be classes of points, but not necessarily to be built up out of points. As we saw in Chapter 1, spatial regions, as continuous wholes, were traditionally treated as wholes given before their parts. The extensions of terms, on the other hand, are wholes made up of discrete parts (the things to which the term correctly applies). These are wholes given after their parts, whether they are given extensionally by aggregating, or listing, a determinate membership, or whether they are given intensionally by a condition which sets a limit or boundary within the totality of actual or possible objects. In the former case the identity of the whole is wholly dependent on that of its parts. In the latter case, the identity of the whole is dependent on the boundary-setting condition, which determines its constitution out of parts. Thus, what happens is that with the use of spatial diagrams in logic, 'class' comes to stand for any whole which has 'parts' (or members) and the relations dealt with are subsumed under the relations of part to whole regardless of very different kinds of relations in which a 'whole' might stand to its 'parts'. Here we have so far distinguished wholes given before from wholes given after their parts, and within the latter have distinguished wholes which depend for their identity on that of their parts from those whose parts depend on the condition which determines the identity of the whole.

Traditional logic thus tended to blur, rather than to clarify the notion of a 'class'. It is within metaphysics and epistemology that the debates which affect the precise understanding of classes as the extensions of terms are to be found. In other words, the debate concerning the relation between potential and actual infinity is, as one might have expected, ultimately a metaphysical, rather than a logical, debate. It was in Aristotle that classical finitism found its first clear exposition and it is therefore to the way in which Aristotle interconnects his views on metaphysics, knowledge and logic that one naturally first looks for the basis of a possible defence of the view that a potential infinity does not necessarily presuppose an actual infinity, or that the possibility of mathematical theorizing, whether about triangles or numbers, does not have to rest on the existence of a completed, determinate totality of triangles or numbers.

2 Aristotle on Knowledge of Universals

It is clear that the claim 'all As are B' entails that any A must also be B. In other words,

All As are B. x is A. Therefore, x is B.

is a valid form of inference. The question is whether it also entails the existence of a completed or actual totality of As, every one of which is B. This would be the case if 'all As are B' were read as equivalent to the conjunction

'a_1 is B & a_2 is B & a_3 is B & ...'

where $a_1, a_2, a_3, \ldots$ is a listing of all the As. For on this reading 'all As are B' is merely a shorthand summary of a list of singular statements, and so it can have a well-defined content only if the list is regarded as complete. This requires that the totality of A's must be a completed totality. Since a potentially infinite totality is, by definition, always incomplete, generalization over such a totality would seem to be impossible without presuming an underlying actual totality (it is only our listing which is incomplete, not the totality being listed).

But this reading of 'all As are B' gives rise to the puzzle (discussed by Aristotle in the opening chapters of his *Posterior Analytics*, and reiterated by Mill (1843)) that *either* knowledge of its truth is humanly impossible (because no finite human can survey the totality of As) *or* it can be known, but this knowledge has no useful applications, since, in order for 'all As are B' to be known, all instances of As to which it might be applied must have already been surveyed. Aristotle's solution is the theory of essence and demonstration propounded in subsequent chapters of the *Posterior Analytics*. This resolution then serves to support his solution of Zeno's paradoxes in that it provides for generalization over potentially infinite, or other indeterminate domains, without the presumption of any underlying actual, completed or determinate totality.

Aristotle's response to the question of how it is that knowledge of the form 'all As are B' can be both possible and useful is indicated in the following two passages:

For one should not argue in the way in which some people attempt to solve it: Do you or don't you know of every pair that it is even? And when you said Yes, they brought forward some pair of which you did not know that it was, nor therefore that it was even. For they solve it by denying that people know of every pair that it is even, but only of anything of which they know that it is a pair. (1984, *Posterior Analytics*, 71a32–35).

Even if you prove of each triangle either by one or by different demonstrations that each has two right angles – separately of the equilateral and the scalene and the isosceles – you do not yet know of the triangle that (it has) two right angles, except in the sophistic fashion, nor (do you know it) of triangle universally, not even if there is no other triangle apart from these. For you do not know it (of the triangle) as triangle, nor even of every triangle (except in respect of number; but not of every one in respect of art, even if there is none of which you do not know it). (74a26–34)

In the first passage Aristotle implies that there can be no solution if it is insisted that 'all As are B' be read as an indefinitely long conjunction of instances. Usually it will not be possible to inspect all instances, in which case a claim to know that every A is B must either be unjustified or be read as meaning merely 'all As I know of are B'. In other words, the belief that every A is B may be useful when we apply it to instances of which we previously had no knowledge (as when making predictions), but it can only be a belief and cannot rank as knowledge. The knowledge that every A we have ever encountered has been B will be knowledge, but will not be useful in that it merely summarizes past experience and does not cover new instances. (This point will be familiar from the discussions of induction which have followed in the wake of Hume, who was simply spelling out the consequences of a thorough-going denial of the Aristotelian tradition.)

In the second passage, Aristotle is making several points which he discusses in more detail elsewhere. He distinguishes between knowing of each and every kind of triangle separately (supposing this to be possible) that its internal angles add up to two right angles (knowing that every A is B by knowing that a_1 is B and a_2 is B and a_3 is B . . .) and knowing of triangles universally that their internal angles add up to two right angles (knowing that all (possible) As are B by knowing what it is to be a triangle and that this entails having internal angles which add up to two right angles). The point is that knowledge *that* the internal angles of a particular kind of triangle add up to two right angles is not knowledge of *why*, in virtue of being a triangle, its internal angles *must* add up to two right angles. It follows from this that Aristotle would equally distinguish between knowing of each individual triangle that its internal angles add up to two right angles and knowing of triangles universally (*qua* triangles) that this is (and must be) the case.

To know of each and every A separately that it is B is only theoretically possible if the membership of the class of As is completely determinate. Possession of this knowledge would require knowing (of the existence of) each A and knowing that the As known (to exist) are all the As that there are (will be or have been). Whereas to know of As universally that they are B does not require knowledge of what particular As there are; it requires knowledge of what it is to be an A (of the form or essence of As) together with a demonstration that whatever is A must, in virtue of its being A, be B. So knowing why all (possible) As are (and must be) B does not involve knowledge of what As (if any) there are, and hence does not involve knowledge of the form 'a_1 is B and a_2 is B and a_3 is B ...'.

So Aristotle's resolution of the problem of how knowledge of 'all As are B' can be both possible and useful is to distinguish between two ways in which something of this form might be known. It might be known factually or theoretically; it might be known merely *that* it is the case, or it might be known *why* it must be the case. It is the factual reading (which Leibniz called 'existential' – see table 2.1) which treats 'all As are B' as a conjunction of its instances. It is in this case that knowledge is only possible where the As form a surveyable (and hence finite) domain, and is such that when acquired it cannot be applied to new instances, as all the instances have, *ex hypothesi*, been inspected. Its use is then to summarize existing knowledge. The theoretical reading (which Leibniz called 'essential' – see table 2.1) treats 'all As are B' as a statement about all possible As. Knowledge that this is the case can only be grounded in an understanding of what it is to be an A.

It might be thought that Aristotle is here merely appealing to a difference in the way that a certain kind of fact may be known and hence that his distinctions will have little bearing on the question of whether any potential infinity presupposes an actual infinity. But, although the distinction is made in terms of knowledge, it is not merely seen as a distinction concerning the *way* in which something is known. It also concerns what it is that is known. This becomes clearer when we realize that Aristotle is *not* claiming (as Leibniz did) that every statement of the form 'all As are B' can be given both readings. To see why, consider the following example:

'All the objects in my pencil box are less than
16 cm × 7 cm × 1 cm.'   (1)

I can know this without even looking inside the box since these
are its external dimensions and anything which could possibly be
inside it must have smaller dimensions. So when I open the box and
see that it contains a ruler I can conclude, without looking further
that it must be less than 16 cm long and is therefore most likely to be
a 15 cm ruler. Even if it had contained a snake, neatly coiled, I
could conclude that the diameter of its body must be less than 1 cm,
and could, with a bit of calculation, put an upper limit on its
uncoiled length (given assumptions about the minimum possible
diameter of a snake). In both these cases I would be applying
theoretical knowledge to particular cases. Before opening the box I
would have only universal knowledge, since I would not know
exactly what objects, if any, the box contains. When I opened the
box I might find that it contained all new pencils and that

'All the objects in my pencil box are 15 cm × 5 mm × 5 mm.' (2)

is true. (2) would be a factually established universal statement,
which merely summarizes my findings about each particular item.
Since there is no reason why a pencil box should contain only
pencils, although it may happen to do so, (2) will be an accidental
truth, knowable only by establishing the facts, whereas (1) could
either be established factually, or, once the dimensions of the box
are known, could be established theoretically, based on an understanding of what-it-is-to-be an object inside this box with its given
dimensions (and bringing in all sorts of background knowledge
about the behaviour of physical objects and presuming that
containers which are small on the outside, but contain a whole
realm within, exist only in fairy tales). But there will not be many
theoretically interesting generalizations to be made about items in a
pencil box, simply because these items (ballpoint, pencil, pen,
paper clips, paper tissue, dead fly, penknife, piece of string) need
have nothing beyond a certain maximum size in common. Being
contained in a given box is an accidental, as opposed to an essential,
characteristic of such things. That is to say each one can be removed

from the box without altering its distinctive characteristics, without losing its identity as a pencil, or whatever.

Generalizations concerning such accidentally collected things will thus, for the most part, only ever be factual generalizations established by inspecting each member of the collection. Theoretical knowledge can only be expected where it is suspected that there is some homogeneity amongst the items forming the collection and that the kind of homogeneity there is has some bearing on what is being said about them (we can expect theoretical knowledge of the maximum size of objects in a container of given dimensions, but not of their colour). In particular, if the things we are talking about are thought to be members of a single 'kind', one might hope to be able to give an account of what-it-is-to-be a thing of that kind (an account of essence) on which to base theoretical, universal knowledge of things of that kind.

So it will not, in general, be possible to have useful theoretical knowledge of 'all A s are B' if this is a generalization over an accidental collection, i.e. if there is no account to be given or sought of what it is to be an A. Here no determinate bounds can be set on what are possible A s, so the A s have to be treated case by case. Moreover, it can only be actual A s which can be considered as forming the class over which to generalize, and the generalization will be treated as the conjunction 'a_1 is A & a_2 is A & a_3 is A & ...', for each actual a_1. It follows that the truth conditions of such a factual statement require the extension of A to have a determinate membership and to be a completed actual totality. Its truth or falsity could only be known to a human being if the A s form a finite, surveyable collection.

In the case where A *does* determine a species or kind and theoretical knowledge is possible, this possibility does not rest on any assumption about whether actual A s do or do not form a totality with a determinate membership, for the possibility of theoretical knowledge of A s rests on the *possible* A s sharing a common nature. In this case the generalization is not over actual A s, but over all possible A s and the relevant extension of A would be the class of all possible A s and this class is not required to have a determinate membership or to form a completed, actual totality.

Thus, although there may be some cases where it is possible to have both factual and theoretical knowledge that 'all A s are B' (a) it will not be the same proposition which is known in two ways, but two different propositions, reflecting different readings of the single form of words, which are known, and (b) in general it will be the case that one reading rather than the other is appropriate, depending on whether the A s form an aggregate of heterogeneous things or a class of things similar in kind. The kind of knowledge possible thus depends on the sort of totality that the A s form, which in turn depends on the sort of term that A is. It is this point on which the viability of the classical finitist position turns.

If, in generalizing over the natural numbers, we are generalizing over a collection of possibly diverse things, which just happen to have the property of being numbers, then, in order for this knowledge to be thought possible at all (for there to be facts of this form to be known), the natural numbers would have to be regarded as forming a complete totality with a determinate membership. But if we are generalizing over a kind of thing, if there is an account to be given of what-it-is-to-be a natural number, then there is a basis for theoretical knowledge of all possible natural numbers which does not presuppose that these form a completed totality. The possibility of the indefinite continuation of the number sequence must thus be seen to be grounded in an account of what it is to be a natural number rather than being grounded in the existence of an actually infinite totality. Similarly for the points on a line.

The defence of classical finitism, then, rests on maintaining two distinctions between kinds of wholes (a) between continuous wholes, given before their parts, and discrete wholes, given after their parts, and (b) between accidental collections (or aggregates) of heterogeneous things, and kinds of things which share a common nature, or essence. It is only discrete collections which can be counted and thus have a number put on them as a measure of their size, and it is clear that if a class is to have a finite number of members it must be a collection with a determinate membership. Thus the class of all possible A s, as a class with an indeterminate membership, cannot be assigned a number. So neither the class of possible men nor the class of natural numbers can be assigned a finite number.

Further, if the notion of an infinite number is conceived by analogy with finite numbers, as a measure of the size of a discrete collection, it is clear that if it were to have any legitimate employment it could only be as applied to non-finite classes whose membership could none the less be considered to be determinate, i.e. to be classes of actual things. This is why it is important to the classical finitist to argue for the finitude of the actual, physical universe and to resist the (platonist) suggestion that there is a universe of actual but non-physical entities which is independent of human thought. Given the finitude of the universe of actual things, all non-finite classes will be classes whose membership is indeterminate, i.e. classes to which the notion of number does not apply. Thus, even with the admission of potentially infinite classes there will be no basis for introducing infinite numbers.

But, as we saw above, these distinctions between kinds of wholes are blurred by the formal logic tradition inherited from the scholastics and continued into the nineteenth century. This was, in part, because it was not possible to distinguish formally between accidental and essential generalizations, or between cases where the extension of a term is an accidental collection of heterogeneous things and where it forms a kind. This would require reference to the meaning of the term and a knowledge of how to distinguish between the accidental and the essential properties of things. This was not thought of as something which could in general be determined a priori. Rather, it was part of the project of science, conceived on an Aristotelian pattern, to gain knowledge of natures or essences.

Aristotle recognized that we reason both from premises which express factual knowledge and from premises which express theoretical knowledge. His basic syllogistic logic is thus throughout open to dual interpretations (his modal logic might be seen as a study of the relation between the two interpretations). The formal rules of his syllogistic reasoning are rules valid under both interpretations and are thus valid whether the extensions of terms have a determinate or an indeterminate membership, are heterogeneous aggregates whose identity is wholly dependent on their membership, or are homogeneous kinds whose identity is given by an account of what it is to be a thing of that kind, which gives an account of the nature of any possible member.

3 Nominalism and Extensionalism

The distinction between accidental collections, or aggregates, and kinds can provide the Aristotelian with a basis for defending the classical finitist position, if it is accepted that mathematical entities, such as natural numbers, points on a line, etc. form kinds. But this defence then rests on a metaphysical assumption that things do indeed fall into kinds and that there are natures or essences to be discovered, that there is an account to be given of what it is to be a thing of a given kind, an account which can form the basis of theoretical, as opposed to merely factual knowledge of things of that kind.

This assumption has been denied by nominalists, from Ockham onwards. They maintain that the world is a world of particulars only. All classification is an imposition by the human mind (whether this is as a product of human nature, as in Ockham (1974, chs 1, 15–17) or of human convention in defining words, as in Hobbes (1665) pt. 1, ch. 2)), and is grounded in human purposes and perceptions, not in the nature of things. All classes are therefore classes of (heterogeneous) particulars accidentally grouped. Consequently there can only be accidental generalizations over such classes and the only form of objective knowledge (knowledge of things as they are in themselves, independent of human projects and perceptions) is (mere) factual knowledge; there can be no objective grounding for claims about what is possible or necessary. Those cases where it seems that we have knowledge of the form 'all *A*s are *B*' without having inspected all the *A*s are merely cases where the proposition in question is in fact a logical truth, something which is true in virtue of the way in which the terms *A* and *B* have been defined. It is something which could be logically derived from a full definition of these terms. If this is the case, then the only limitation on what is possible is that imposed by the logical requirement of non-contradiction (nothing can be both *A* and *not-A* at the same time). (Note, that, as is so characterized, the nominalist position is essentially a realist one, it is made out by insisting on a distinction between how things are and how we think about them, and between what can be the objective content of a statement and the means by which it may come to be known or be believed to be true.)

It is thus part of the nominalist position to claim that mathematics, which provides the most obvious examples of generalizations which can be known to be true without inspection of the instances to which they apply, is merely the result of working out the logical consequences of definitions (consists merely of analytic truths). Clearly such analytic truths can be established without knowing what As there are, or indeed whether there are any. So it would be misleading to represent the content of what is known as the conjunction 'a_1 is B & a_2 is B & a_3 is B & ...'. The knowledge can only relate to the universe and actual particulars, and is knowledge that *if* a thing satisfies the definition of A, then it will also satisfy the definition of B, i.e. for every object x, if x is A, then x is B. This generalization itself can only be treated as an accidental generalization, i.e. as equivalent to the conjunction '(if o_1 is A, then o_1 is B) & (if o_2 is A, then o_2 is B) & ...' which lists all o_i in the universe of actual objects (whatever that may be taken to be). This reading of theoretically established universal truths commits the nominalist to thinking of the universe of actual objects as itself a completed, actual totality with a determinate membership. Whether he is also committed to thinking of the extension of A as determinate depends on the exact form the nominalism takes (i.e. on the nature of the particulars admitted into the actual universe) and on the reading accepted for non-analytic, factual universal statements.

· If classes are admitted they will have to be treated as accidental collections, determined by their members, and hence as having a determinate membership. In this case the only available reading of universal statements will be an extensionalist one. But assuming that common terms have determinate extensions presents problems. If the universe of particulars includes human beings, who can change their marital status without losing their identity, then the extension of 'bachelor' would be determinate only at a given time and would fluctuate over a period of time. For this reason it is less problematic to make the reading of universal statements independent of classes and to treat all cases alike as saying 'for every object x, if x is A, then x is B', giving this generalization an extensional reading. However, in the case of factual generalizations this does raise the question of whether the truth of a conditional statement 'if c is A, then c is B' has to depend on there being either some general connection between being A and being B, or some particular connection, dependent òn the nature of c, between cs being A and cs being B. In the former case the analysis of the universal statement would become directly circular, and in the latter case the rejection of kinds and their essences would have been replaced only by embracing individual essences or natures. On the other hand, if conditional statements can be true without any further connection between the properties involved, they have to be given a material reading 'it is not the case that c is A and c is not B'. In the case of universal statements this takes us back to construing 'all As are B' as saying that there is no actual A which is not also B, which does presume that each object in the universe is determinately either A or not, and so presumes that there is a determinate collection of things which are A.

The exact specification of the universe of particulars is a problem which the nominalist has to address. But the general structure of the metaphysical framework and its implications for logical theory are in many respects independent of the universe adopted. The structure is that set out most explicitly and thoroughly in Wittgenstein (1922). However, the specification of the universe of particulars does affect the nominalist account of mathematics. For example, if numbers, classes, points, etc. are admitted into the universe of actual particulars, then universal generalizations over these will be represented as statements of the form 'for every x, if x is a number, then . . .' and so on. Since mathematical entities, unlike physical ones are presumed neither to change over time, nor to come into and go out of existence, the list of entities which are numbers can be assumed to have a fixed, determinate membership. Thus the potential infinity of the number series could only be grounded in an actual infinity of numbers.

If numbers, classes, points, etc. are, as abstract, non-physical entities, excluded from the universe of particulars, then, if mathematical statements are to be represented as having any objective, cognitive content at all, they will have to be reconstrued as statements about the world of non-mathematical particulars. Terms referring to individual numbers, classes or points will have to be assumed to be merely part of a convenient (but sometimes misleading) abbreviated language used to talk about the physical world. Whenever such terms are used they do not really refer to objects, as their grammar would suggest. In principle they could be eliminated in favour of a much longer statement in which the only objects referred to were concrete particulars. It was to provide a demonstration that, and of how, this translation could be realized that Whitehead and Russell (1910–13) embarked on the project of writing *Principia Mathematica*. On this approach, the infinity of the number series is only assured if the actual universe contains infinitely many individuals. This would seem to make the infinity of the number series a contingent truth. How is this to be reconciled with the existence of a proof which appears to establish that there can be no largest natural number, and to establish this as an analytic truth? The answer is that this proof has to be reassessed in order to see what exactly it does prove. It must be taken as proving not that no last number exists, but that no last number can be named. Any time we compute an upper bound we can always add one to the number so computed and thus have a larger number. This however, does not mean that we can compute, and therefore write down, names for all the numbers, if the universe is finite in both space and time.

An extreme nominalist position can thus provide a motivation for strict finitism (a position which will none the less encounter the problems associated with space and time, discussed in chapter 1). If the nominalist is going to countenance the infinite at all, it will have to be an actual, not merely a potential infinity, whether this is within the actual, physical universe, or in a universe which includes abstract objects.

4 Toward an Algebra of Classes

With the seventeenth-century overthrow of Aristotelian essential-ism in science, nominalism of one form or another has been the dominant metaphysical position. The extensionalism which is the most natural consequence of this position certainly simplifies the situation as regards classes as the extensions of terms. These are always to be thought of as determinate collections of actual objects. A collection can then only be potentially infinite (such that we will never come to an end when inspecting its members or when listing them) if it is actually infinite.

Now although it was said that the core of Aristotle's syllogistic theory was neutral between existential and essential interpretations, this neutrality was bought at the price of adopting as valid only those forms of inference which remain valid for reasoning to establish both factual and theoretical conclusions, rejecting any which might be valid under one interpretation but not the other. There are syllogistic forms which Aristotle did not regard as valid, but which were included by scholastic logicians as valid. This is because they did not, as Aristotle did, draw any firm distinction between accounts of essence and the definitions or meanings of words (instead we get discussions of the relation between real and nominal essence and whether they can be distinguished). It tended to be assumed that every term has both an extension and an intension (meaning) and thus that every sentence capable of entering into a syllogism could be given either an extensional or an intensional reading, which is possible only if the account of intensional readings is cut free from its anchor in Aristotle's epistemology and metaphysics.

The source of the difference lies in the status of negative terms. Aristotle does not treat negative terms as proper terms, terms fully on a par with others. It might be thought that 'all swans are black' is logically equivalent to (says the same thing as) 'all non-black things are non-swans', and indeed they are equivalent if either (a) it is assumed that we are talking about the totality of actual things which are swans or not and black or not (for the extension of $S(not\text{-}B)$ is empty if, and only if, the extension of $(not\text{-}B)(not\text{-}not\text{-}S)$ is empty, since the extension of $not\text{-}not\text{-}S$ must then be the same as that of S); or (b) it is assumed that we are talking about a relation between intensions, where the intension of $not\text{-}A$ is defined by assuming the equivalence of 'c is not A' and 'c is $not\text{-}A$', and also that the negation of the negation of a proposition p is equivalent to p (for then to say that $S(not\text{-}B)$ is a contradictory concept will be the same as saying that $(not\text{-}B)(not\text{-}not\text{-}S)$ is a contradictory concept, since the intension of $not\text{-}not\text{-}S$ will be the same as that of S). But if we were thinking of this as a generalization over swans as a kind, giving it a formal (essential) or theoretical reading, then we immediately encounter an asymmetry. For neither non-black things, nor non-swans constitute kinds. Indeed, in general the things not of a given kind will be a very heterogeneous lot. So negative terms can only have a place when the reasoning concerns accidental collections. In theoretical reasoning the two statements will not be equivalent because 'all non-black things are non-swans' will not be susceptible of a theoretical reading at all. Having done away with kinds, however, the nominalist can feel free to treat negative terms on a par with any others.

This has already been tacitly illustrated in Leibniz algebraic symbolism and it is the move which paves the way for an algebra of classes. This still allows for both extensional and intentional readings, for the nominalist needs to retain intensions as the meanings of terms for his account of analytic truths. But it rules out the equation of intensions with accounts of essence (unless, like Leibniz, one assumes that each individual has its own essence). However, even from an extensionalist point of view, the introduction of negative terms is not without problems. These can be seen by returning to the spatial representations of classes employed by Euler and Venn. Here the extension of a term is represented by a circle (a bounded spatial region) but the negation of that term will then be represented by an unbounded region – the rest of the universe. The problems encountered when trying to think about the universe as a whole have already been mentioned in chapter 1. If the universe is a whole, it is of a very special kind. Diagrammatically and algebraically the problem is not very difficult to resolve. Into the diagram we merely have to introduce a representation of the universe (figure 2.3). Algebraically we just need a symbol, such as

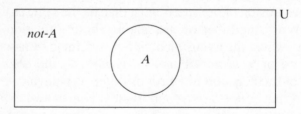

Figure 2.3

U, to stand for the universe, a symbol '∅' to stand for the empty or null class together with the rules that for any term A, $A + not\text{-}A = U$ and $A(not\text{-}A) = ∅$, where '+' is an operation on classes such that the members of $A + B$ (the union of A and B) are the members of A together with the members of B, and AB (the intersection of A and B) contains just those things which belong to both A and B. It then becomes natural to think of *not-A* as $U - A$. The rules can be readily translated from the spatial representation to give the familiar Boolean algebra for classes:

$$A + A = A \qquad\qquad AA = A$$
$$A(not\text{-}A) = \varnothing \qquad\qquad A + not\text{-}A = U \qquad \varnothing \neq U$$
$$A + B = B + A \qquad\qquad AB = BA$$
$$(A + B)B = B \qquad\qquad AB + B = B$$
$$A(C + B) = AC + BC \qquad\qquad A + BC = (A + B)(A + C)$$

The problems lie in the interpretation of these extra elements introduced to smooth up the algebra. If they are also to be thought of as standing for classes (if this is to be an algebra of classes and not just a symbolism which may assist in reasoning about classes), then U and $\varnothing$ must stand for classes.

De Morgan circumvented the problems associated with the universe by introducing the notion of a universe of discourse. This is not the totality of all possible objects but the whole of some definite category of things, considered to be the things under discussion in a given context. This then has the effect of limiting negation to negation within the universe of discourse. Such a device will work, and is justifiable within a framework in which genera and species are accepted, but it must appear *ad hoc* when they are not. From Aristotle's point of view a term represented as a negation within a universe of discourse would not strictly be a negative term, for A and *not-A* will now be contraries rather than contradictories. Their relation will be like that of 'even number' and 'odd number' – every (natural) number is either even or odd, but there are many things which, not being numbers, are neither even nor odd.

It is easy enough to give sense to $\varnothing$ via the intensional reading of terms – it is the extension of a contradictory term, one under which nothing falls. But it is not easy to make sense of it on the under-standing of classes as aggregates of objects. On the other hand, the understanding of *not-A* makes most sense when thinking about classes as aggregates of objects and is more problematic from the intensional point of view. Moreover, Leibniz regarded his algebra as being capable throughout of both extensional and intensional interpretations. Of the two kinds of spatial representation, Euler's works by representing classes as classes of actual objects and the relations which may actually obtain between these, whereas Venn represents possibilities by spatial regions. All the representations push inexorably toward a mathematically elegant Boolean algebra as a calculus of classes and as a mathematization and simplification of traditional syllogistic logic. But they do nothing to clarify the notion of a class as the extension of a term and they leave the status of the universe and the empty, or null class problematic. In the handling of negative terms there is a vacillation between a strict nominalism and something closer to an Aristotelian position. The difficulties mentioned here present problems which have to be resolved if there is to be a fully coherent notion of 'class' as the extension of a term within a nominalistic framework. They play a crucial role in the attempts by Frege and Russell to clarify the situation and to develop a more powerful extensional logic.

5 Classical Finitism in Retreat?

If it is thought that Aristotelian essentialism is banned by taking modern science seriously, does it then follow that classical finitism must collapse with it? It certainly cannot flourish within a nominalist framework, for here, as we have seen, Pascal's point must stand – any potential infinity presupposes an actual infinity. The nominalist will have to opt either for strict finitism or for the existence of an actual infinity. This, of course, does not commit him to saying that we can know anything of, or make any mathematical sense out of, the actual infinite. Again, like Pascal, he might take the existence of the absolute infinite as expressing a limitation on our capacity for knowledge.

The crucial question is whether the distinction that the classical finitist requires, between theoretical and factual knowledge as a distinction not merely between ways of knowing, but also between the things known, can be grounded in anything other than an Aristotelian metaphysics. Although the issue about kinds and essences is the most obvious point of disagreement between Aristotle and the nominalist, this itself turns on a more fundamental issue, one which survives the overthrow of Aristotelian science. This is the issue about the nature of particulars or individuals. For the nominalist the world is a world of particulars. These exist in their independent particularity prior to any thought about them, prior to any classification imposed on them. Particular existence precedes any universals or concepts. For Aristotle, on the other hand, every individual substance was a combination of form and matter. Form gives discrete being, as a thing of a given kind, to formless matter, whereas matter gives individual distinctness to different instantiations of the same form. From this point of view there can be no objects, no particulars, without universals, without conceptualization.

The alternative to nominalism which survives is one which rejects the total primacy of particular over universal, and which insists that all objects, as possible objects of thought, discourse or knowledge must be conceptualized in some way, must be classified as objects of some kind or other. One cannot thus talk about objective knowledge of the world of particulars as they are in themselves and independent of all conceptualization, because the very notion of a particular is one which makes sense (is possible) only in the context of conceptualization. A re-emphasis on the interdependence of concepts and objects as the objects of knowledge or discourse has emerged, in various rather different forms, from the philosophy of science in recent years (from Quine (1953), with the denial of the distinction between analytic and synthetic truths, to Kuhn (1962), Feyerabend (1975), and beyond).

This coordinacy of the notions of concept and object is also strongly insisted upon in the work of Frege, for instance, and is an integral part of the structure on which he relies in his attempts to disentangle the relations between the notions of class, number and concept. In chapter 7 it will be possible to consider Frege's position in more detail and to judge the extent to which it either shows the classical finitist position to be untenable or provides for the possibility of non-Aristotelian justification for it.

3

Permutations, Combinations and Infinite Cardinalities

We have seen that within a nominalist framework, Pascal's point stands; every potential infinite presupposes an actual infinite. But we have also seen how this is linked to an extensionalist reading of universally quantified statements and that on this reading the truth of such statements is possible only if either it is based on knowledge of the meanings of the terms involved, or it is a generalization over a finite, surveyable collection. For the majority of generalizations of the 'all swans are white' sort, neither of these will be the case. This commits the nominalist, such as Pascal (and Hume), to saying that knowledge is not possible here. But since we do have evidence in the form of observations of particular instances, the focus of epistemological attention shifts to inductive reasoning and to the question of whether, even if knowledge of universal truths is not possible, we can none the less get estimates of probabilities. Pascal himself contributed to the mathematical study of probabilities and it is within the mathematical disciplines which developed around the estimation of chances and probabilities that there emerges a distinctively mathematical, combinatorial mode of thinking about possibilities and totalities.

Here collections, or aggregates, are made up of discrete objects (are wholes given after their parts) and their membership is considered to be determinate with the identity of the whole firmly dependent on that of its members. Moreover, the totalities are not restricted to being finite, for the epistemological challenge faced by a theory of probability applicable to inductive inference is to be able to cover cases where the things over which the generalization is to be made form a non-surveyable domain whose exact membership is unknown and which may be infinite. The basic combinatorial approach to chances, possibilities and probabilities is, however, rooted in thinking about finite collections; its extension is a matter of trying to treat the infinite by analogy with the finite.

1 Finite Permutations and Combinations

Given a bag containing four balls, two of which are white and two black, the chance of selecting two black balls when asked to take out two balls can be computed by determining the total number of possible ways of selecting two balls out of four and determining how many of these would be selections of two black balls. Clearly there is only one way to select two black balls, because in this case there are only two of them, but there are six possible ways of selecting two balls (figure 3.1) There is thus a 1 in 6 chance of

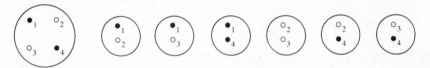

Figure 3.1

selecting two black balls and a 4 in 6 chance of getting one white ball and one black. In general there are $n!/p!(n-p)!$ possible ways of selecting p things from a collection of n things. If we wanted to work out the chance of selecting one black and one white ball selecting a black first, we would need to consider the number of possible ways of taking out two balls in a given order. Each of our previous pairs could have been selected in two different orders (two possible permutations), so there are now 12 possibilities, four of which would be selections of black followed by white. So the chance is 1 in 3. Generally there are $n!/(n-p)!$ differently ordered selections (permutations) of p things which can be made from a collection of n things.

This is the basic idea behind the theory of permutations and combinations which is applied in complex ways to deal with games of chance and with evaluating probabilities. From the present standpoint, however, our interest is in the way in which totalities are treated here. In order to compute chances and probabilities in these sorts of cases it is necessary to be able to calculate the numbers of possible outcomes. This can be done via calculations of numbers of permutations and/or combinations, i.e. of possible ways of selecting from a given set which is presumed to have a fixed, finite and heterogeneous membership. The set considered must be a determinate collection of discrete things, a whole with which it is possible to associate a number. Moreover the concern is not with making generalizations over this class but with making statements about another class – the class of all possible selections of a given kind from this set. If we have a collection of n things, then the total number of possible ways of selecting subcollections, of whatever size, from it will be 2^n because, for each of the n elements, we can either include it in our subcollection or not, and n such two-way choices determine a subcollection. So for two elements we get four possibilities.

$$
\begin{array}{ccccc}
a & 0 & 0 & 1 & 1 \\
b & 0 & 1 & 0 & 1 \\
& \varnothing & \{b\} & \{a\} & \{a, b\}
\end{array}
$$

where $\varnothing$ is the empty set, 0 means the element gets excluded and 1 means that it gets included.

Here we are dealing with finite, discrete, determinate sets and the possible permutations, combinations or selections of their elements are discussed with a view to establishing the probability of a certain kind of selection, given that each possibility is equiprobable, i.e. that it is simply a matter of chance as to which is selected, or that the selection is random. In the case of throwing a die, for example, there are six possible outcomes which are equipossible if the die is not biased.

But in real life many of the occurrences for which we would like probability estimates are specified as members of classes whose membership is indeterminate and where we do not know how to go about enumerating a background set of possible outcomes within which this represents a selection. Thus we do not know how to enumerate the possibilities or whether all relevant possibilities are equally probable. We may have stable mortality statistics, but not only do we not know the number of possible causes of death, but it is not clear that the causes of death form a class with a determinate membership – one which could sensibly be expected to be assigned a number. What we do in practice is to take figures from what we judge to be a sufficiently large 'random' sample and extrapolate. The question, first taken up by Jakob Bernoulli (1713), is whether this procedure can be given any theoretical justification and whether there are limits to the degree of confidence to be placed in such extrapolations.

2 Probabilities as Limits of Infinite Sequences

Starting from an imagined chance set-up on which repeated trials can be made, Bernoulli presumes that there is a constant but unknown chance p of a particular outcome S for any given trial. If n trials have been made, a proportion of s_n of S outcomes will have occurred. He proved that the probability that in a sequence of n trials the proportion s_n of S outcomes is as close as desired to p increases with n, i.e. the probability that

$$|p - s_n| < \varepsilon$$

tends to 1 as n becomes indefinitely large. And for a given error ε he shows how to compute an n such that the probability of getting

$$|p - s_n| < \varepsilon$$

is greater than a given probability $1 - \delta$.

This theorem, if its conditions can be fulfilled, gives an assurance that the larger the sample taken, the more likely it will be that extrapolations based on it are reliable, provided that the sample is an arbitrary or random one. But the theorem does raise difficult problems of interpretation. What is meant by saying that there is a constant unknown chance p of a particular outcome S, e.g. death as the result of heart failure, when the number of diseases or other possible causes of death is unlimited or unknown?

When thinking about dice or pebbles in a bag there were a determinate number of possible outcomes (the range of possibilities). If these are each equipossible, then we expect that on a random sequence of trials the outcomes will be more or less equally divided between all the possible outcomes, i.e. the chance p of the outcome of a given trial being S, assumed in Bernoulli's theorem, can be calculated by reference to the range of possibilities. In these cases the proportion of S outcomes in a sequence of n trials (the relative frequency of S outcomes in the sequence) can be seen as an approximation to a value which is determined by the range of possibilities.

In trying to extend this to get estimates of things like the chance of dying as a result of heart failure, where the range of possibilities is unlimited, we cannot think of p as being determined in the same way. Instead we might try adopting a more extensional approach. We assume that the class of human deaths has a determinate membership and that each death either is or is not a result of heart failure. Then there must be a determinate proportion p of this total number of deaths which were the result of heart failure. In this case the value that the relative frequency of deaths from heart failure in any given random sample approximates to is the relative frequency of such deaths in the total class of human deaths. What Bernoulli's theorem would then say is that, even though we cannot survey the total class of human deaths, as our sample gets larger (assuming it to be random) we get a better estimate of the relative frequency over the total. This way of thinking is on a par with that according to which we say that there is an actual infinite which we can know to exist (because we can construct potentially infinite sequences) but about which we can know nothing more. Similarly, it might be thought that we can know that there must be a determinate distribution of S outcomes in the totality of trials of a given chance set-up, but if we cannot come to an end of the trials we cannot know this distribution, we can only estimate it on the basis of finite sequences of observed trials.

This makes sense so long as the class of actual human deaths is finite, but not if it is infinite, for there is no way of assigning a numerical value to the proportion p (division by infinity does not make sense). And it should be noted that the application of Bernoulli's theorem involves thinking of the class of trials as potentially infinite (and hence presupposing it to be actually infinite). On the other hand, when the actual outcomes are restricted to a finite number of possibilities the extension to an infinite case is quite possible. This can be seen by considering the natural numbers. We cannot estimate the proportion of natural numbers which are even numbers by comparing the number of natural numbers with the number of even numbers because there are infinitely many of each and in fact just as many even numbers as natural numbers. Each natural number can be uniquely 'named' by an even number.

1 2 3 4 5 ...
2 4 6 8 10 ...

But since, of the infinitely many possible choices of number, we know that each must be either odd or even and that the cases are equipossible, we can none the less say that the ratio of even numbers to numbers is 1:2, or the chance of getting an even number is 1 in 2, and the larger the finite set we get the more likely it is that we will get half of it as even numbers.

There is, however, an alternative way of thinking about the infinite case. This involves weakening the analogy with the finite case and treating Bernoulli's theorem as saying that in a genuinely chance, or random, set-up, the sequence of values s_n for the proportion of S outcomes on a sequence of n converges and it is this convergence itself that gives sense to thinking that there is a definite chance p of a particular outcome being S for any given trial. The sequence does not converge because there is an independently determined value for p which its members approximate, i.e. it does not converge to a pre-existent limit. It is because it converges that it makes sense to talk of p, and to say that it has a limit means no more than to say that it converges. This question of how to interpret the existence of limits of infinite sequences is one to which we shall return. The point is that the availability of this means of making sense of talking of the limit is easily combined with thinking that the limit, none the less, has an independent existence and thus justifies a supposition that the infinite case really is analogous to the finite case, we just have to use different methods because we can neither survey nor put numbers on infinite classes.

Interpretations of Bernoulli's theorem and of probabilities are controversial. They are mentioned here to show how this combinatorial mode of investigation of possibilities, originally linked to the investigation of chances and probabilities, is extended beyond finite and well-bounded classes. This is done by treating indefinitely large classes as potentially infinite, but against a background which requires that the potential infinity of trials or samples be grounded in the supposition of an actual infinity, and thus legitimates the supposition that the distinction between finite and infinite collections is contingent upon the limitations of our cognitive faculties. It thus pushes inevitably in the direction of an extensional and combinatorial, distinctively mathematical, treatment of collections and their arbitrary subcollections.

3 Infinite Cardinalities

However, any fully combinatorial approach to infinite sets and to a theory of possible selections from them appears to be blocked by the property that an infinite set can in some sense be said to have just as many members as a proper subset of itself (e.g. there are just as many even numbers as there are natural numbers). This is true not only for the natural numbers but also for the points on a line.

In figure 3.2, suppose AB is twice as long as CD. A line drawn from a point x on AB to E will intersect CD in a single point x′. Thus to every point on AB there corresponds exactly one point on CD. Further, if perpendiculars are dropped from CD onto AB, each point of CD will be projected onto a point in C′D′, which is a proper segment of AB. Combining the two mappings, every point of AB is mapped onto a point in C′D′. So there must be just as many points in C′D′ as in AB even though AB is twice as long. The number of points in a line does not correlate with, or in any way determine, the length of the line.

Dedekind took this property as being that which distinguishes finite from infinite sets: an infinite set can be put into one–one correspondence with a proper part of itself, no finite set can. Since

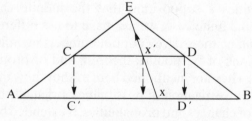

Figure 3.2

we can make sense of putting infinite sets into one–one correspondence with each other there is a sense in which, provided we consider them to be actually infinite, we can think of two infinite sets as containing 'the same number' of elements. But this 'number' is not something which could ever be assigned by or reached by counting. So if we were to think of actually infinite, determinate sets as having a number of elements, we would have to allow that what cannot be counted must be allowed, on some occasions, to have a number.

When two sets (whether finite or infinite) can be put into one–one correspondence with each other they are then said to have the same power or cardinal number. For finite sets this also separates having a number from being counted, but still preserves all the normal results, for the cardinal number assigned to a finite set will be the same as its ordinal number – the one it would be assigned by counting its members. But in the infinite case it will mean that an infinite set may have the same cardinality as a proper part of itself. This is, to say the least, paradoxical if one is thinking of number as a measure of a plurality, of the size of a discrete collection. The size of the whole should always be greater than that of any of its proper parts. There must surely be more natural numbers than there are even numbers (indeed twice as many), and more points on a line two units long than on a line one unit long. It seems that the part–whole relation and notions of magnitude based on it have to part company with the application of numbers. The notion of number arising from the existence of a one–one correspondence (cardinality or power) cannot be seen in any straightforward way as a measure of the size of an infinite collection.

This consideration by itself would suggest that the fundamental distinction is that between finite and infinite, and that infinite sets are without number not only because they cannot be exhaustively counted, but also because even the notion of cardinal number, as a measure of the size or numerosity, can get no grip here. And this would be the correct conclusion to have drawn had it not been for Cantor's discovery that not all infinite sets have the same cardinality, i.e. there are infinite sets which cannot be put into one–one correspondence with each other.

First one can note that in the case of any finite set A, the set of all possible subsets of A is always larger than A, indeed if A contains

n elements, its power set, the set of all its subsets, contains 2^n elements. One can then consider whether this result holds only for finite sets, or whether it is true in general. It is clear that given any set A, its power set must contain at least as many elements as A itself because for each $a \in A$, $\{a\}$ belongs to $P(A)$. Suppose then that we had some way of mapping each element of $P(A)$ onto a unique element of A, e.g. by a function f such that

(a) for $s \in P(A)$, $f(s) \in A$ and
(b) if $s, t \in P(A)$ and $s \neq t$ then $f(s) \neq f(t)$.

For each $s \in P(A)$, either $f(s) \in s$ or not, since $f(s) \in A$ and s is a subset of A. So there must be a subset r of A which consists of those elements $f(s)$ such that $f(s)$ does not belong to s. r must itself be mapped onto an element of A and we have either $f(r) \in r$ or not.

If $f(r) \in r$, then $f(r)$ belongs to a collection consisting wholly of elements which do not belong to the sets to which they are related by f, which is a contradiction.

Buf if $f(r)$ does not belong to r, then $f(r)$ must be in r, since r contains all elements of A which do not belong to the sets with which they are correlated.

So in either case there is a contradiction, and the original assumption that $P(A)$ can be put in a one–one correspondence with A must be false.

In this argument A could be either finite or infinite so the result holds good even for infinite sets. So, for example, the set of all subsets of the natural numbers has a greater cardinality or power than the set of natural numbers. By analogy with the finite case, if the number $\aleph_0$ is assigned as the number of natural numbers, the number of possible subsets of this set will be $2^{\aleph_0}$. This suggests that one should, after all, if one is going to admit actually infinite sets and consider taking arbitrary subsets of them, distinguish between orders of infinity, or between orders of infinite magnitude.

4   The Finitist Response

Here we would appear to have a route via which the introduction of actually infinite sets and of infinite cardinal 'numbers' may be justified. But does the classical finitist have to accept this justification? It is fairly clear that he cannot object to the combinatorial approach to finite totalities, provided these are given as discrete totalities whose membership is fully determinate. But he might well object to the presumption that this approach can be extended from such totalities to discrete totalities whose membership is either indeterminate or potentially infinite. For, to make the extension, such indefinite totalities are treated as if they were potentially infinite. And the finitist might well insist that there is a distinction to be drawn between indefinite and potentially infinite. For all we know the actual extension of a term for a kind might be finite. Classes whose membership is indefinite, and which are specified by necessary and sufficient conditions for membership (by drawing a boundary within a given universe of discrete objects), are to be distinguished from those which are potentially infinite. Potential infinity can only be judged relative to a principle of generation, one that can be shown to have no end or to be continuable indefinitely. Potentially infinite classes are thus also always given with a natural order, the order of generation. This is not true of indefinite classes.

On this ground then, the classical finitist might say that there is no basis for treating indefinite classes as potentially infinite and therefore no ground for treating them as actually infinite either. To treat them by the methods appropriate to determinate totalities is justified only on the highly realist assumptions of the extensionalist who wants to say that the indeterminacy is only relative to our knowledge; in reality every well-defined term has a fully determinate extension. Given that the finitist has other reasons for rejecting this assumption, he will resist the pressure to treat chances relating to indeterminate totalities by analogy with the estimation of chances in gaming situations, and will seek alternative interpretations of what is to be meant by probability and of methods of estimating it in more normal, everyday situations.

But Cantor's theorem applies to potentially infinite totalities, such as that of the natural numbers. Does the classical finitist have, none the less, to admit that in this case we have to admit the existence of different orders of infinity? To see that he does not, we first have to note the difference in character between the set of all natural numbers, N, and its power set, $P(N)$. The natural numbers form a potentially infinite totality, and indeed are the paradigm of such a totality. Can $P(N)$ be supplied with a principle of generation and hence qualify as a potentially infinite set? Consider the full binary tree shown in figure 3.3.

Each level of construction of this tree can be indexed by a natural number. If a 0 at level n indicates that n does not belong to the set and 1 indicates that n does belong to the set, then each path on the tree can be seen as representing a subset of the natural numbers. The leftmost path is $\varnothing$, the empty set, and the rightmost one is the full set N. A path given by $\langle 0\ 1\ 0\ 0\ 1\ 1 \ldots \rangle$ translates into the set $\{1, 4, 5 \ldots\}$.

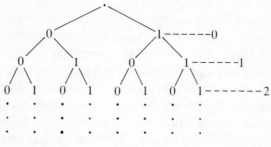

Figure 3.3

So the power set of N, or of any other potentially infinite set, can be seen to be potentially infinite. What then of the supposition that one might define, or that there might be, a one–one correspondence between $P(N)$ and N? Here the finitist has, even without Cantor's theorem, grounds for thinking that no such correspondence is possible. $P(N)$ is for him a potentially infinite set whose members are themselves given only by a potentially infinite definition and which are thus never fully given. $P(N)$ is not a set composed of fully defined, discrete objects. At any stage of construction only a finite part of the definition of any subset will have been given, a part which it will have in common with infinitely many other subsets. The idea of having a one–one correspondence between two sets requires that to each element of one set there corresponds exactly one member of the other. But there is, on extensional grounds, never any way of dealing with exactly one member of $P(N)$. Thus $P(N)$ is not a totality of discrete objects and does not qualify either for being put into purely extensional one–one correspondences or hence for having a cardinal number assigned to it. It is a different kind of potentially infinite set from N and is generated in a different way.

Thus the finitist can agree that there can be no one–one correspondence between N and $P(N)$ without drawing the conclusion that this means that there must be distinct orders of infinity leading to higher infinite cardinalities. Instead he will have distinct kinds of potentially infinite set, some whose members are discrete and fully defined, others where the members themselves are never fully actual and hence never fully distinct – given any finite path in the binary tree it can be extended in more than one way to yield two (and more) distinct members of $P(N)$.

The combinatorial approach to the investigation of possibilities, chances and probabilities in a nominalist and extensionalist framework contributes to the development of a mathematical concept of a class as a collection or aggregate whose identity is determined wholly by its membership, and to the development of the concept of cardinal as distinct from ordinal number (arising from the distinction between combinations and permutations) and thus to a legitimation of infinite cardinal numbers. But there is nothing here which will compel the classical finitist to shift his ground. The interpretation of probability is widely recognized as being controversial and there are many who reject the relative frequency approach on grounds which have nothing to do with issues relating to infinity. If he is already resisting extensionalism, the finitist will resist the indiscriminate application of a combinatorial approach to totalities, selections from them and other operations on them. The greater pressure comes from the other mathematical context in which infinite series are extensively employed, namely, in the quest for an arithmetical representation of the continuum.

4

Numbering the Continuum

There are two senses in which the continuum can be said to have been numbered. (1) The (linear) continuum has been replaced by, or is even in some circumstances identified with, a set of numbers – the real numbers. (2) The continuum has been assigned a cardinal number $2^{\aleph_0}$, i.e. sense has been given to the question 'How many points are there in a line?' and a partial answer given. In the light of what was said in chapter 3 it is clear that (2) could not have come about without (1) or something like it. The continuum had to be represented by a set of points with a determinate membership before it could be assigned a number.

How could this transition have come about? How were the paradoxes of the infinite overcome? So far we have seen that the classical finitist can stand his ground, admitting only the notion of the potentially infinite, without thereby being committed to the existence of any actual infinite provided that (a) he distinguishes between continuous and discrete wholes, between wholes given prior to their parts (where the identity of the parts depends crucially on that of the whole of which they are part) and wholes given after their parts (where the identity of the whole is determined by that of its parts); (b) that he insists on the distinction between essential and accidental generalization, or at least between extensional and non-extensional readings of 'all As are B'; and (c) that he insists on the distinction between the indefinite and the potentially infinite.

Moreover, even if one admits that every potential infinite pre-supposes an actual infinite, this still does not overcome the apparent contradiction involved in thinking of a continuum as made up of points. The actual infinite, even if it is metaphysically inevitable, does not thereby become a possible object of knowledge, or a contradiction-free and therefore usable mathematical concept. Pascal indeed used the actual infinite as a foil, as a means of proving the existence of a being, knowledge and understanding of which transcends all human rational capacities. We can know of its existence but cannot comprehend it. This is not without significance for, as we shall see, Cantor too was obliged to admit a notion of the absolutely infinite, which he also associated with God and which had to be placed outside the range of mathematical computation and comprehension.

In this situation the classical finitist has a strong case. He is at least in a good position to engage in metaphysical and epistemological arguments with his opponent and has the upper hand epistemologically, where it would seem that the actual infinite can play no significant role. But his position will be changed radically if (a) it can be shown that a coherent conception of the continuum as an infinite collection of points is after all possible, and (b) that the actual infinite plays a significant role in the mathematics which is used in and is necessary to the natural sciences, and physics in particular.

1 The Algebraization of Geometry

The pressures which brought down the edifice of classical, Aristotelian finitism did indeed come from within mathematics and physics. Ultimately they derive from the demand for a numerical, practically applicable handling of continuous magnitudes and in particular of continuous change (including, of course, motion). With hindsight it can be seen that the crucial moves had already been made by Descartes in his *La Géométrie*, where he argues that Euclid-style definitions of geometric figures should be replaced by definitions given in the form of algebraic equations. From this the notion of a function rapidly followed in the work of Leibniz and Newton, and it is the subsequent development of this concept (which is all-important to the mathematical physicist) which apparently dictates the punctualization of the continuum. But at the same time it introduces a new, specifically mathematical conception of totality (set or class) – a whole given neither before nor after its parts, whose membership is to be regarded as determinate, generalization over which must be treated as extensional but non-accidental. In other words, there arise mathematical conceptions of totality which cross-cut prior philosophical distinctions and which primarily inform Cantorian set theory and subsequent axiomatizations.

The logicist programme for providing a rigorous foundation for analysis superimposed the logical notion of class on these mathematical conceptions in a way which conceals their distinctive character (for indeed the logicist claims that there is here no distinction). But, in the face of Zeno's paradoxes, a condition of the possibility of treating the continuum as a totality of points, without absurdity, is the emergence of new ways of thinking about totalities, new ways of conceptualizing and reasoning about continuous wholes which synthesize the traditionally distinct notions of continuous and discrete wholes. This opposition had to be transcended in the production of any such synthesis.

It will, therefore be necessary to sketch the course of this synthesis and the emergence of new ways of conceptualizing totalities. This can be no more than an impressionistic sketch, for the history here is complex and technical (any mathematically and philosophically rigorous treatment would require many volumes). Those wanting more rigour and/or more detail are referred to the suggestions for further reading at the end of the book.

It should come as no surprise to find that mathematical, rather than philosophical, considerations are those which pose the real challenge to the classical finitist. It is important to locate this challenge more precisely than is frequently the case, for it is only in this way that we can come to see exactly what sort of sense is made of the actual infinite within mathematics and so to assess the wider implications of its mathematical use. It is customary to treat the invention of infinitesimal calculus as marking the occasion of the really significant intrusion of the actual infinite into mathematics. While it is true that the calculus was introduced (and perhaps could only have been introduced) in a philosophical climate of metaphysical acceptance of the infinite (a climate of rational theology), it is not the mere introduction of methods of differentiation and integration which dictates the move either to a point continuum or to an actual infinity.

The original introduction of the operations of differentiation and integration was geometrical. As geometrically grounded they can (as the later work of Weierstrass and others showed) make do with traditional geometrical concepts of continuity, the potential infinity of points of division, potentially infinite sequences and the notion of a limit, even if when loosely described they appear to involve infinitesimal magnitudes and actually infinite division. Differentiation can be pictured as giving the gradient of the tangent to a curve $y = f(x)$ at a given point (x, y) by treating it as the gradient of the line from (x, y) to $(x + \delta, f(x + \delta))$ when δ is infinitely small (figure 4.1). (A condition for this to be defined is that one should get the same result by approaching from the right, i.e. by considering lines from $(x - \delta, f(x - \delta))$ to (x, y).)

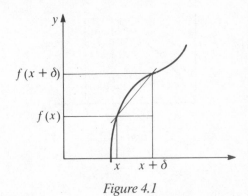

Figure 4.1

This would treat the gradient as the ratio of two infinitely small quantities $f(x + \delta) - f(x)$, and δ, but raises the awkward question of how an infinitely small quantity differs from 0, and of how one can divide by such a quantity and distinguish the result from division by any other infinitely small quantity. Paradoxical conclusions quickly follow as Berkeley pointed out in his criticism of Newton's use of calculus (Berkeley, 1734). These difficulties are avoided by treating the gradient of the tangent as the limit of an infinite sequence of ever closer approximations each of which is a ratio between finite lengths, i.e. as

$$\lim_{\delta \to 0} \frac{f(x + \delta) - f(x)}{\delta}$$

Similarly the definite integral can be pictured as giving the area under a given section of a curve $y = f(x)$. This area can be approximated by chopping it up into rectangles and adding their areas together. The narrower these rectangles are, the better the approximation is. The areas might thus be thought of as the sum of infinitely many, infinitesimally thin rectangles which still somehow manage to have a non-zero area. But the area can also be defined as a limit approached by summing over successively thinner, but still finite, rectangles (figure 4.2).

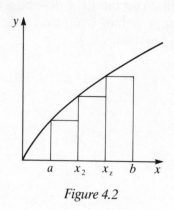

Figure 4.2

$$\int_a^b f(x) = \lim_{n \to \infty} \Sigma((x_1 - a)f(a) + (x_2 - x_1)f(x_1) + \ldots + (b - x_n)f(x_n))$$

The idea of a limit, the limit of an infinite sequence of ever closer approximations to a given quantity, thus supplants the infinitesimal and it would initially seem that such sequences need only be regarded as potentially infinite. For the idea of an infinite sequence of closer approximations to a given limit is already present in Zeno and the classical finitist could get away with potentially infinite sequences there. He can continue to do so in this case provided that he sticks to differentiation and integration as geometrically picturable and interpretable operations. For here the limits to which an approximation is sought are already geometrically defined independently of any sequences of approximation to them. The gradient of the tangent to the curve is geometrically given by the ratio AB/XB, i.e. this ratio exists, the only problem is to put a number on it. Similarly, any closed plane figure is presumed to have an area, even though its measurement may be problematic. The infinite sequences involved are then just the familiar potentially infinite sequences associated with continued, ever finer division.

Restricted to this almost purely geometric form, differentiation and integration are merely modifications of existing geometrical techniques (Archimedes methods of exhaustion and of indivisibles, and various methods for constructing tangents to curves). What then, was the essential novelty? For there is no doubt that the methods of infinitesimal calculus were new and powerful, and moreover that they were perceived, both at the time of their introduction and since, as involving the infinite in mathematics in a way which did not occur with the geometer's recognition of the infinite divisibility of a continuous magnitude.

What is new is the fact that they form part of a calculus. It is the context of introduction which makes the difference. These things are not merely conceptualized as limits of infinite sequences of approximations, but are associated with methods of computing values of limits of such sequences. As such they are part of, and indeed central to, the motivation of the algebraization of geometry. It is the algebraic representation of the (potential) infinite already inherent in standard geometrical practice that gives it a new and problematic, because number-like, status. The algebraic representation gives us at least the appearance of being able to calculate with the infinite and with infinitesimals. If we write

$$\lim_{n \to \infty}$$

for example, it is tempting to read '$n \to \infty$' in the same way as '$n \to 1000$' and thus as if ∞ were some value that n might actually attain, even though the geometrically guided use of the whole limit expression neither warrants nor requires this. Moreover, although the algebraic notation and its associated operations were initially introduced as linked with an intended geometrical interpretation, they soon take on a life of their own, going beyond what is geometrically representable or picturable. The question of what sort of sense is to be made of operations and expressions which have their origin solely in the algebraic, symbolic representation then becomes urgent and there is a pressure to give arithmetical, numerical interpretations primacy over geometrical interpretations and hence for a rigorously and independently founded arithmetical–numerical representation of the continuum.

It was Descartes who first systematically introduced algebraic methods into geometry, insisting that the objects of geometry, geometrical figures and curves, should be defined not in the manner of Euclid, but by algebraic (polynomial) equations. He thought of such an equation as giving the law according to which a point would have to move in order to generate a curve. Thus for example a circle, centre (a, b) and radius r, is given by the equation

$$(x - a)^2 + (y - b)^2 = r^2$$

Its circumference is traced out by a point (x, y) which moves in such a way as always to satisfy this equation (figure 4.3). There are here several important moves away from the classical tradition of Euclid and/or Aristotle. They were not all initiated by Descartes but were first brought together by him.

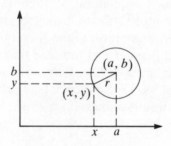

Figure 4.3

In the first place, Descartes is using variables, symbols such as 'x' and 'y' to stand for quantities which change. Their immediate interpretation is a geometric one; they stand for distances or lengths along a given axis from the point where they intersect (the origin). But Descartes also quite explicitly treats these lengths as themselves representing any other continuous magnitude which one might happen to be interested in. Thus change of temperature with time may be pictured geometrically by taking lengths on one axis as representing temperatures and on the other axis as representing times. Moreover the same goes for areas and volumes, these too may be represented, as continuous magnitudes, by line lengths. This is a departure from earlier practice where it was presumed that if a and b are line lengths then $a \times b$ stands for an area. Because Descartes is prepared to allow that any continuous magnitude can be represented by a line length he can allow that there must be a line length to represent $a \times b$. In this way he makes line lengths closely resemble numbers in that multiplication and division are now operations defined over line lengths, whereas previously only addition and subtraction had seemed to make sense. This allows Descartes to make geometric sense of expressions such as x^5 which would otherwise have to have been thought of as the 'volume' of a five-dimensional cube. It is this step which is crucial to being able to regard a polynomial equation, such as

$$y = a_1 x^5 + a_2 x^4 + a_3 x^3 + a_4 x^2 + a_5 x + a_6$$

as defining a curve in a two-dimensional space. Descartes thus assumes that all continuous magnitudes and all ratios between them (whether they are commensurable or not) can be represented by lengths. In this way the theory of ratios and proportions between continuous magnitudes was swiftly turned into an 'arithmetic' in which ratios are treated as numbers of a new kind. These ratios would include not only those between commensurable magnitudes, but also those between incommensurable magnitudes, i.e. ratios known not to be expressible as ratios between whole numbers, such as $\sqrt{2}$ and π.

Secondly, the focus of geometric attention is turned away from closed figures to continuous paths, whether forming closed curves or not and on their algebraic characterization – the characterization of a 'motion' which will generate the curve. This means that one is no longer dealing with continuous wholes as wholes which are bounded and limited and in this way given before their parts. Instead they are treated as generated wholes which may be potentially infinite but which are given by the algebraically expressed law constraining and determining their generation.

It is important that the early development of algebraic, analytic geometry was closely, and indeed almost inseparably, bound to the development of mathematically expressed theories of mechanics. It is this which means that the most common way of thinking of the

relation between a curve and the algebraic expression which
defines it is by thinking of the expression as a law of a generating
motion. Calculus itself was developed with a view to providing a
quantitative treatment of change, and rates of change. It is thus in
terms of motion that limits are interpreted and understood. Thus
Newton wrote:

> Perhaps it may be objected that there is no ultimate propor-
> tion of evanescent quantities: because the proportion before
> the quantities have vanished, is not the ultimate: and when
> they are vanished, is none. But by the same argument, it may
> be alleged that a body arriving at a certain place, and there
> stopping, has no ultimate velocity; because the velocity before
> the body came to that place is not its ultimate velocity: when it
> has arrived it is none. But the answer is easy; for by the
> ultimate velocity is meant that with which the body is moved,
> neither before it arrives at its last place and the motion ceases
> nor after, but at the very instant it arrives: that is, that velocity
> with which the body arrives at its last place, and with which the
> motion ceases. And in like manner, by the ultimate ratio of
> evanescent quantities is to be understood that ratio not before
> they vanish, nor afterwards, but with which they vanish. ....
> There is a limit which the velocity at the end of the motion may
> obtain, but not exceed. This is the ultimate velocity. And there
> is the like limit in all quantities and proportions that begin and
> cease to be. And since such limits are certain and definite, to
> determine the same is a problem strictly geometrical. (New-
> ton, 1934, pp. 38–9)

In this context differentiation is (a) defined as an algebraic
operation, and (b) interpreted as giving the rate of change of one
quantity (represented on the y-axis) with respect to another
(represented on the x-axis) at a point. Thus if

$$y = ax^2 + bx + c, \qquad \frac{dy}{dx} = 2ax + b$$

and the 'rate' of change of y with respect to x does not have to be
worked out for each point separately; it is given by a new equation.

Integration is also defined as an algebraic operation and as the inverse of differentiation; so

$$\int 2ax + b \, dx = ax^2 + bx + k$$

Moreover, their definition as algebraic operations allows for repeated applications (even though these may not always have any natural physical interpretation). If dy/dx gives the rate of change of y (position) with respect to x (time), i.e. velocity, then d^2y/dx^2 gives rate of change of velocity with respect to time, i.e. acceleration. This makes it possible to talk of and put values on instantaneous positions, velocities and accelerations whilst also having equations which characterize the ways in which they are changing. The algebraic characterization of both motions and geometric curves thus marks an enormous increase in descriptive power. An equation is a source of information about any point one chooses (and in this sense is an infinite description of all points) which is also a characterization of the whole which is not built up from information about points.

So besides the potentially infinite, discretely generated sequence of the natural numbers, there is now, in addition, the conception of a line which is continuously generated, in accordance with a law, and which does not involve constructing a point from those which preceded it, but merely ensuring that a constant, complex ratio, expressed algebraically, is preserved. Regarded in this way we can imagine the construction of a graph in the following way: a point can be construed as carrying out motions away from the origin in the direction of both the x- and y-axes simultaneously. Suppose that

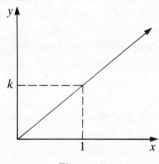

Figure 4.4

motion along the x-axis is uniform, continuous and has unit velocity. If motion in the direction of the y-axis is also uniform, continuous and has velocity k then the resulting graph will be a straight line whose gradient is k and the equation of the motion will be $y = kx$ (both motions being referred to the same equably flowing time). This is shown in figure 4.4.

Suppose now that motion in the y direction is initially v, but is uniformly decelerated (acceleration $-a$), then $x = t$ and $y = \frac{1}{2}[v + (v - at)]t$, i.e. $y = vx - \frac{1}{2}ax^2$, and what we get is the path of a projectile (a parabola) as shown in figure 4.5. The composition of the non-uniform motion with the uniform motion has the effect of 'bending up' the straight line which forms the x-axis.

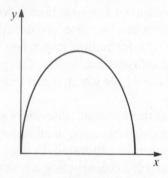

Figure 4.5

But just how non-uniform might the motion in the y-direction get? If $y = \sin x$ we have a wave function. And wave functions can themselves be superimposed on one another to give quite complex patterns – as when the ripples sent out by two or more pebbles meet. So we might get a motion which is 'oscillatory' in a complicated way. Daniel Bernoulli (1700–1782), when approaching the problem of how to write down an equation for the motion of a vibrating string said:

My conclusion is that all sonorous bodies include an infinity of sounds with a corresponding infinity of regular vibrations. . . . Each kind multiplies an infinite number of times to accord to each interval between two nodes an infinite number of curves, such that each point starts and achieves at the same instant, these vibrations, while following the theory of Mr Taylor, each interval should assume the form of the companion of the cycloid extremely elongated. (Manheim, 1964, p. 41)

The equation given to reflect the superposition of this infinity of curves is

$$y = \alpha\sin\frac{\pi x}{a} + \beta\sin\frac{2\pi x}{a} + \gamma\sin\frac{3\pi x}{a} + \delta\sin\frac{4\pi x}{a} + \ldots$$

The idea that one could use such a superposition of 'wave' functions to represent, algebraically, a given curve over a given interval proved to be crucial both for the development of the concept of a function and of set theory.

At the time at which Bernoulli was writing, it was presumed that two functions which coincide over an interval will coincide everywhere and that any algebraic equation in which y is given as a function of x is geometrically representable, whilst not all geometrically drawable curves are algebraically representable. The initial problem was precisely that of finding analytic, algebraic expressions (laws) to characterize given curves or 'motions'. And there was disagreement between D'Alembert who equated a function with its algebraic expression, and Euler who identified a function with its graph. The efforts to generalize Bernoulli's results and to find the conditions under which an infinite trigonometric series actually represents a given function led eventually to the theory of point sets and provided the stimulus for Cantor's introduction of ordinal numbers.

Fourier gave a precise statement of the generalized problem: given an arbitrary function $f(x)$, find the coefficients a_n and b_n such that the equation

$$f(x) = \frac{a_0}{2} + \sum_{n=1}^{\infty} (a_n \cos nx + b_n \sin nx)$$

shall be an identity over a prescribed interval of the x-axis. The very statement of this problem marks a shift firmly in the direction of (a) equating a function with its graph, for Fourier's arbitrary function means an arbitrarily drawn function, (b) treating algebraic expressions as capable of representing (piecewise if necessary, i.e. using different representing functions for different intervals) every drawable geometric curve.

In the course of investigation of this general problem posed by Fourier, a number of 'pathological' functions were discovered, functions which were algebraically expressed in terms of infinite sums but whose 'graph' is unpicturable. To see how these can arise we have to note that, given the indefinite divisibility of the continuum, there is no limit to the number of oscillations that can be packed into a given interval (their frequency), and given its unbounded nature there are no upper limits which can be placed on the amplitude of such oscillations. Functions which are expressed as complicated wave functions will, however, always be continuous (see figure 4.6). A discontinuous function is one which 'jumps' at one or more points (see figure 4.7).

If there is no limit to the frequency with which motion in the *y* direction might oscillate, then might it oscillate with an infinite frequency? That is, might it be the case that no matter how small an

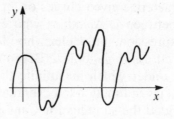

Figure 4.6

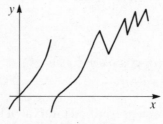

Figure 4.7

interval we take there will be an oscillation contained within it (i.e. the graph will have changed direction) in this interval? This is what would appear to be the case with Weierstrass's everywhere continuous but nowhere differentiable function

$$f(x) = \sum_{n=0}^{\infty} b^n \cos(a^n x)$$

where a is an odd integer greater than 1, b is a positive constant less than 1 and ab is greater than 1. It can be shown that for any point x_0 the difference quotients (giving 'gradients') approaching from the right and from the left have a different sign no matter how close one gets to x_0. So the function, though continuous, is not differentiable at any point. Riemann provided an example of a function which has infinitely many discontinuities between any two limits but which is none the less integrable.

$$f(x) = \frac{(x)}{1} + \frac{(2x)}{4} + \frac{(3x)}{9} + \ldots = \sum_{n=1}^{\infty} \frac{(nx)}{n^2}$$

where

(x) = the excess of x over the nearest integer
(x) = 0 if x is midway between two integers

so

$$-\tfrac{1}{2} < x < \tfrac{1}{2}$$

$f(x)$ is convergent for all values of x, but is discontinuous for all x of the form $p/2n$, where p and n are relatively prime. Thus $f(x)$ is discontinuous an infinite number of times in every arbitrarily small interval. But $f(x)$ is not too wild; the number of jumps larger than a given s is always finite. So it is possible to chop the regions into small enough bits so that within each bit the jumps are smaller than s, and so to get successive approximations of the 'area under' $f(x)$.

With such pathological functions we see the power of the algebraic symbolism and symbolic operations to outstrip geometric intuition. These functions are not picturable and thus disrupt the assumption, based on picturable cases, that continuity, integrability and differentiability go together. But more than this, the infinitely dense packing of either continuous or discontinuous oscillations compels recognition of a complex structure in the apparently simple, homogeneous, equably smooth flowing linear continuum. Such functions have the power to introduce divisions in the continuum, not one at a time, but in an unpicturably infinite density all at once, as it were. It should be noted that this is done with functions which are themselves defined using limits of infinite series.

It is here that we have the ground of the undermining of the classical finitist position. Functions had been conceived in inseparable association with their graphs – the 'paths' traced by points moving in accordance with an algebraically expressed law. But when that law dictates a 'motion' which involves infinitely frequent oscillations, or infinitely frequent jumps, it is a path which can no longer be geometrically traced either in the mind's eye or on paper. But if the law can be written and by this means rationally investigated, the graph of the function must be presumed, in some sense, to exist and to be a totality of points over which our only hold is now algebraic. These points are the members of the set of values of a function $f(x)$ for each x considered as a numerical argument. Thus it becomes necessary to think of the original, smoothly continuous line as itself a set of points, each indexed by a number, and which has an unimaginably, because infinitely, complex order structure.

Sets so conceived are actually infinite totalities, given neither before nor after their points. Not before, because the points are no longer points of potential division successively generated and not after because the totality is not defined by reference to characteristics of points (for points are in themselves identical to one another). Indeed, if we are thinking geometrically, a function does not uniquely define a set of points in the plane, for the set of points which constitute its graph is dependent on the unit chosen. To take a very simple example, consider our parabola given by the function $y = ax - bx^2$. Under two different choices of unit it will define two different sets of points (figure 4.8). (The inner curve results from doubling the unit on both axes, i.e. is the graph of $y'' = ax'' - bx''^2$ with $x'' = 2x$, $y'' = 2y$.)

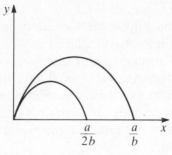

Figure 4.8

Geometrically, then, it is only relative to a unit, or measure, that a function can be correlated with a set of points. What the algebraic expression does is to replace the old geometric definitions of figures in the sense that these are definitions characterizing a certain kind of spatial structure or configuration, which can be realized on all sorts of different scales. When we think algebraically and think in terms of sets of numbers, rather than sets of points, then it is clear that the set of numerical values of a function is unique. The numbers here serve as a means of specifying a structure which, in some cases we have no other means of specifying, a structure which can be instantiated in many different ways depending on the way in which points, lengths or other continuous magnitudes are assigned numbers (measured). In the case of the function neither law nor graph takes precedence, for the graph, the path, is the geometrical/ mathematical object of study, but it is given as such only via the function which defines it. Complex structures are the wholes with which the mathematician is concerned and which his algebraic notation also suggests are wholes composed of indivisible points – sets of points indexed by numbers. If the function is well defined (can be shown to have a unique value for every argument) then membership of the set of points constituting its graph (relative to a given measure) must be determinate. But since the path may be indefinitely long, the sets of points may be indeterminate in the sense of being unbounded. Generalizations over such sets of points are grounded in the function of which they are the graph, not in the characteristics of points. It is the structure of points, the relations between them, which are important.

The pathological functions finally showed the power of algebra to outstrip geometrical intuition. Geometry and motion could no longer be relied upon to provide the basis for analysis; it could not function as the background against which its operations were to be interpreted and tested. Analysis would have to become algebraically autonomous. This was the motivation behind arithmetization. There was a need to provide an account of continuity, differentiability and integrability in numerical terms, terms not drawing on geometrical intuition.

2 The Arithmetization of Analysis

But how was a purely arithmetic theory of limits to be constructed? Starting from the positive and negative integers it is possible to define positive and negative rational numbers (or fractions) as ratios between integers, and to give rules for their addition, subtraction, multiplication and division. But the rational numbers, although densely ordered (between any two rational numbers there is always a further rational number), do not form a continuum. There are more points on a line than can, after selection of a unit, be represented by rational numbers. This is the traditional problem of the existence of incommensurable magnitudes. Lines of, for example, length $\sqrt{2}$ can be geometrically constructed and proved not to be representable by any rational number. So geometrically the limit of a sequence of rational approximations to $\sqrt{2}$ is known to exist. But if geometrical intuition is to be dispensed with, then it cannot be presumed that the limit of such a sequence of rational numbers exists; rather such limits have to be introduced by means of definitions which do not presuppose their existence. From the purely arithmetic basis there are as yet only rational numbers and infinite sequences of rational numbers, some of which converge and some of which do not. The aim is to arrive at an arithmetically defined/constructed set of numbers which could adequately represent (by indexing) the points on a line.

The intuitive link between the conceptions 'point' and 'real number' is clear in Dedekind's way of defining real numbers. After noting that the straight line L is indefinitely richer in point individuals that the domain R of rational numbers is in number individuals, because there are infinitely many points in the straight line that correspond to no rational number, he says:

> If now, as is our desire, we try to follow up arithmetically all phenomena in the straight line, the domain of rational numbers is insufficient and it becomes absolutely necessary that the instrument R constructed by the creation of the rational numbers be essentially improved by the generation of new numbers such that the domain of numbers shall gain the same completeness, or as we may say at once, the same *continuity*, as the straight line. (Dedekind, 1963, p. 9)

This led him to ask 'In what does this continuity consist?' His answer is that the essence of continuity lies in the following principle:

> If all the points of the straight line fall into two classes such that every point of the first class lies to the left of every point of the second class, there exists one and only one point which produces this division of all points into two classes, this severing of the straight line into two portions. (Dedekind, 1963, p. 11)

The 'discontinuity' of the rational number sequence is thus seen to lie in the fact that not all cuts in it are produced by rational numbers, where a cut in the rational numbers R is

> any separation of the system R into two classes A_1, A_2 which possesses only *this* characteristic property that every number a_1 in A_1 in less than every number a_2 in A_2. (Dedekind, 1963, pp. 12–13)

Thus if A_1 were the set of all rational numbers less than $\frac{1}{2}$ and A_2 were the set of all rational numbers greater than or equal to $\frac{1}{2}$ this would be a cut produced by the rational number $\frac{1}{2}$. Whereas if A_1 is the set of rationals less than $\sqrt{2}$ and A_2 is the set of rationals greater than $\sqrt{2}$ this would be a cut (every rational number lies on one side or other of $\sqrt{2}$) but it is not produced by a rational number.

In order to create a continuous numerical domain one needs a number system in which all cuts are produced by, or correspond to, numbers of that system. This is achieved by saying that whenever a cut (A_1, A_2) is not produced by any rational number, 'we create a new, an *irrational* number α, which we regard as completely defined by this cut'

> From now on, therefore, to every definite cut there corresponds a definite rational or irrational number, and we regard two numbers as *different* or *unequal* always and only when they correspond to essentially different cuts. (Dedekind, 1963, p. 15)

Moreover, the extended domain of numbers created in this way proves to be closed under the operation of forming cuts. i.e. if one considers forming cuts in the sequence of real numbers, there will always be a real number which produces it. Thus, in this sense, the domain of real numbers can be said to be continuous. In addition, in order to justify his claim to have defined new *numbers*, Dedekind had to show that 'cuts' in the sequence of rational numbers can be added, subtracted, multiplied and divided in such a way that when the cut is produced by a rational number these operations reduce to the familiar operations on rational numbers.

The way in which Dedekind 'creates' his real numbers draws on the combinatorial approach to sets considered in chapter 3. His cuts are amongst the possible selections from the set of rational numbers and the totality of cuts is the totality of arbitrary selections which meet the defining condition for being a cut. It is thus a very non-constructive approach in that it treats cuts as pairs of sets without considering how these sets may themselves be specified. He has to presume that the totality of rational numbers is a fixed, actually infinite totality with a determinate membership; moreover it is a densely but linearly ordered totality (between any two rational numbers there exists another rational number and given any two rational numbers, $a, b, a = b$, or $a < b$, or $a > b$).

If the sets of rational numbers constituting a cut were not thought to have a determinate membership, then the cut itself would not be precisely defined. This approach is non-constructive not only in its use of actually infinite totalities, but also in that the cuts are presumed to exist independently of sequences of approximation or of any means of defining the sets which constitute them. They are introduced by strict analogy with the geometric potential divisibility of the continuum, but once divorced from the geometric analogy, as in all strictness they are supposed to be, they have been cut off from any operation which might 'produce' the cuts and which gives a hold on the notion of division as a potential. The assumption of the existence of limits is replaced by the assumption of the existence of arbitrary selections from a given set.

The alternative route to the creation of real numbers, adopted by Cantor, is in some respects more constructive in its approach. The set of rational numbers R, considered as a representation of the points on a line, is incomplete in the sense that there are lengths, and hence points of division, which can be given a rational approximation of any desired degree of accuracy but can receive no exact numerical representation, i.e. there are infinite convergent sequences of rational numbers which have no rational limit. A sequence $\langle a_n \rangle$ of rational numbers is a *fundamental* sequence if for any positive rational ε there is an integer k such that, $|a_{n+m} - a_n| < \varepsilon$ for any m and all $n > k$. Cantor identified the new (real) numbers with these sequences and showed that one could define an order on them by saying that if $\langle a_n \rangle = b, \langle a_n'' \rangle = b''$, and if for any positive rational ε there is an integer k such that for all $n > k$

$$|a_n - a_n''| < \varepsilon \quad \text{then} \quad b = b''$$
$$a_n - a_n'' > \varepsilon \quad \text{then} \quad b < b''$$
$$a_n - a_n'' > -\varepsilon \quad \text{then} \quad b < b''$$

So two such numbers are equal if the sequences which define them vary by less than any given ε after a finite distance. It follows that given any rational number a, the constant sequence $\langle a \rangle$ (whose limit is a) is such that either $b = a$, or $b < a$, or $b > a$, for any real number b. Arithmetical operations can then be defined for the new numbers

$$b + b' = b'' \quad \text{means that} \quad \lim_{n \to \infty}(a_n + a_n' - a_n'') = 0$$
$$b \cdot b' = b'' \quad \text{means that} \quad \lim_{n \to \infty}(a_n' a_n' - a_n'') = 0$$

So the bs now look and behave like numbers and they incorporate within them a model of the rational numbers. Cantor went on to iterate this process:

　　C is the set of numbers expressed as fundamental sequences of members of *B*

.

.

.

　　L is the set of numbers expressed by fundamental sequences of members of *K*

.

.

.

　　Now the relation between *A* and *B* is different from that between *B* and any subsequent domain reached by this style of definition in that although every *a* belonging to *A* is represented in *B*, there are elements in *B* which have no counterpart in *A*. But for *C* it can be shown that every element of *C* already has a counterpart in *B* (and *B* can be shown to be isomorphic to Dedekind's real numbers). Thus in one sense no new elements are created, however many times the process is reiterated. Yet as sequences the members of *C* are quite distinct from members of *B*, they are sequences of sequences of rationals.

3　Toward Infinite Ordinal Numbers

After selection of a unit of measurement and an origin it can be shown that every point on a line can be indexed by a unique element of *B*, and Cantor postulated that to every element of *B* there corresponds a unique point on the line which it represents. So in *B* we would already seem to have an arithmetical model of the continuum. Why, then, did Cantor go on to iterate the procedure and think it important to distinguish between the members of *B* and subsequently formed 'numbers' even though *B* already contains the limits of all fundamental sequences of elements of *B*? This relates to the motives for wanting an arithmetical representation of the continuum in the first place. It is to try to achieve some representation and understanding of the complexity of its point structure as revealed, for example, by the pathological functions, and to produce a definitive and general answer to Fourier's representation problem. Dedekind's method of introducing real numbers reflects only the basic geometrical intuition of arbitrary divisibility. Cantor is concerned to exhibit the complex fine structure – the order structure – of points on a line. It is to assist with this that he distinguishes between points in terms of the types of series of which they are the limit. He considers a set P of points on the line (considered as indexed by numbers) and gives the following definition.

> By a *limit point of a point set P* I mean a point of the line for which in any neighbourhood of the same, infinitely many points of P are found, whereby it can happen that the (limit) point itself also belongs to the set. By a 'neighbourhood' of a point is understood an interval which contains the point in its interior. Accordingly it is easy to prove that a point set consisting of an infinite number of points always has at least one limit point. (Dauben, 1979, p. 41)

Given any point set P and any other point on the line, it either is or is not a limit point of P. Thus each P has a well-defined set $P^{(1)}$ of limit points, the first derived set of P. If $P = R$ (the set of rationals), $P^{(1)} = B$, the set of all real numbers expressed as limits of fundamental sequences of rationals. But P might be any infinite set, so $P^{(1)}$ might not be B. If $P^{(1)}$ is infinite the operation can be repeated. Either there is some finite n such that $P^{(n)}$ is finite and hence $P^{(n+1)}$ does not exist, or there is not. In the case of R, since iteration always leads to the full set of points of the line, there is clearly no such n. This suggests that if there are sets of points for which the iteration stops at some n, they are not fully continuous although they have a certain non-uniform kind of density. A point p which is expressed as a limit of sequences of sequences of rational numbers is the derived set of a set of points each of which is the limit of a fundamental sequence of rationals.

$$P^{(1)} \,.\quad .\quad .\quad .\quad \ldots\ldots\ldots\ldots\ldots\ldots\quad .\quad .\quad .\quad .$$
$$\{p\} = P^{(2)} \qquad\qquad\qquad\qquad P^{(2)} = \{\lim_{n\to\infty}\langle b_n\rangle\}$$

$P^{(1)}$ is then a set of points which cluster round p. But each of these is itself expressible as a limit of a sequence of rationals. So P is a set of rationals which cluster round p with a particular type of dense ordering. The n, if it exists, for which $P^{(n)}$ becomes finite is then a sort of measure of the 'density' of clustering of rationals in P round a finite number of points. P is not an evenly distributed dense set, but is locally dense in a finite number of places.

But Cantor did not stop here. For if $P^{(n)}$ is not empty for any finite n, then clearly, even if the operation were iterated an infinite number of times, there would still be a set, and Cantor introduced P^∞ to indicate this and $P^{\infty+1}$ to indicate the derived set of P^∞. ∞ and $\infty + 1$ are new infinite, ordinal numbers and they make their first appearance as indices of the iterations of an operation designed to characterize the structural characteristics of sets within a point continuum and hence the possible intricacies of behaviour of functions and their representation by trigonometric series.

The transfinite ordinal numbers thus first come into being as a way of indexing iterations of the operation of forming the derived set of a set of points. They appear to be required by the attempt to characterize the distribution of points in a continuum. As initially introduced, they do not, and were not intended to put a number on the points in the continuum, although this was the question which later came to preoccupy Cantor. Before seeing how this question arises, however, it is worth considering where the developments described above leave the finitist.

4 Conflict with Classical Finitism

The crucial question for the classical finitist is whether the fundamental sequences of rationals, in terms of which the real numbers are introduced, can be considered as potentially infinite sequences, and indeed whether the rational numbers themselves can be considered as a merely potentially infinite totality. Initially the answer in both cases might be thought to be 'Yes', but a closer inspection of what is required to produce an arithmetic point continuum, which can dispense with geometrical intuition whilst at the same time preserving results founded upon it, suggests that this is not the case. The reason for this is that a class cannot be regarded as potentially infinite unless it is thought of as generated by a non-terminating process. So the possible potentially infinite sequences of rational numbers are those which can be generated in some way, either by a law or by a sequence of choices, whether free or constrained. Any such sequence is only ever given as a finite fragment plus a generating process, with the consequence that such a sequence is not uniquely given by its (actual) terms, any more than a species, or a subset, of the natural numbers is given by its members.

Thus the sense in which such a sequence may be said to have a limit must be given by the condition for convergence. A sequence $\langle p_1 \rangle$ of rational numbers is convergent if and only if, for any positive rational number δ, there is a positive integer n such that the absolute value of the difference between p_n and any subsequent term of the sequence is less than δ. To say when a sequence converges is to say what it is for a sequence to have a limit. The limit is not something which is independently given and to which the terms of the sequence can be thought to approximate ever more closely. This means that any extensional statement about such a sequence (i.e. one which is true of any other sequence which is extensionally equivalent to the given sequence) has to be true or false on the basis of a finite amount of information about the sequence. For example, two convergent sequences are said to converge to the same limit if and only if there is some point in the sequences after which the difference between corresponding terms becomes arbitrarily small. If there is no law generating the sequences (they are free choice sequences) then it will not be possible to say that they do or do not converge to the same limit, for all we will have to go on will be finite initial segments of the sequences, segments which give no assurance about how they will continue. Since there are indefinitely many sequences with the same finite initial segment this means that any statement we can make about a lawless sequence (on the basis of knowledge of some finite initial segment of it) will not be true of just that sequence but of indefinitely many – all those which share the particular initial segment.

The totality of all potentially infinite sequences of rationals, or even of all convergent or fundamental sequences of rationals, is thus like the set of all subsets of the natural numbers, in that it is a potentially infinite whole containing members which are never fully determinate, for they themselves are potentially infinite, and always actually finite but necessarily incomplete. This does not yield a point continuum in anything like the sense presumed by Cantor, for its members do not have the precise identity required of points.

When this approach to real numbers is interpreted geometrically we get precisely identifiable points of division corresponding to the rational numbers together with the possibility of focusing on an ever smaller region in order to locate a 'point' indexed by a non-rational number, a possibility which is guaranteed by the dense ordering of the rationals (between any two there is another). But since the process of focusing is not, and never can be, complete, the continuum is never resolved into points, only into ever smaller regions.

For example, consider the set of real numbers between 0 and 1 as given by their binary decimal notation. All such sequences can (as in figure 3.3) be represented by the full binary tree which can be thought of both as the making of successive choices about whether to put 0 or 1 in the nth decimal place and as the making of successive divisions of the unit line. All the points corresponding to an infinite sequence with an initial segment .0001 ... lie in the interval $[\frac{1}{16}, \frac{1}{8}]$. Each addition of a level to the tree chops the line up into smaller bits (figure 4.9).

It is, however, possible to develop a theory of 'real numbers' and functions over them on this basis (see for example Bishop, 1967; Troelstra, 1977). The result is intuitionist, rather than classical, analysis and they are by no means equivalent. But what is important from the point of view of our present investigations is the fact that both approaches can be pursued. This means that the classical finitist does not have to back down. On the other hand, his position has lost much of its plausibility. With the development of the classical theory of real numbers on the foundations suggested by Cantor and Dedekind it has become possible, without obvious

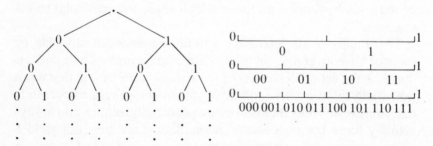

Figure 4.9

incoherency, to think of a continuum as 'made up' of points (strictly as a set of points). Moreover, this view seemed to be required by the way in which functions defined over real numbers are customarily associated with their 'graphs'.

The arguments derived from Zeno's paradoxes against taking a continuum to be made up out of points are circumvented because the 'construction' of the real numbers, which is the initial vehicle for thinking about a point continuum, is not a geometrical construction. It is not a matter of building a continuum by distributing points in space, but of defining the real numbers and showing that these can be ordered so as to be order isomorphic to the points on a line. This means that a one–one correspondence can be established between the real numbers in the interval (0, 1) and any finite line, and between the positive (or positive and negative) real numbers and any infinite line.

It is with the pathological functions that the continuum is actually infinitely divided in a way which is not geometrically picturable. A function which has infinitely many discontinuities between any two limits effects a divorce between the introduction of discontinuities (divisions) and the idea of a successive process. It thus does not support a conception of potentially infinite division, but of actually infinite division. It is the geometrical origin of the notion of a function that suggests that to any well-defined function there corresponds a 'graph' as something which exists as a determinate object of mathematical investigation. It is also the geometric interpretation which is heuristically important in very many of the applications of analysis. But it is the algebraic expression of a function that makes it natural to think of this determinate object as a geometrically determinate, actually infinite set of points.

So the purely arithmetical development of the theory of real numbers (defined as equivalence classes of convergent infinite sequences of rationals) and of functions defined over them does not automatically support a theory of real numbers which can validate the conception of an actually existing infinite point continuum. This is because the infinite sequences of rationals can be treated as either actually or potentially infinite. But when the definition of real numbers as the limits of convergent sequences of rational numbers is given in the context of the intended geometrical interpretation it appears (a) that the actual infinite must be accepted if the real numbers are to provide an arithmetic model for the points in a line in the manner which is implicit in coordinate geometry with its tacit assumption that every point on the x- and y-axes is indexed by a number, and (b) that the geometrical interpretation supports a realist conception of limits of infinite sequences, via the picture of points as limits of sequences of approximation to them, limits which are therefore given independently of the sequences of which they are the limits. It is this which tends to underwrite the conception of infinite convergent sequences of rational numbers as actual, completed, infinities rather than as potential infinities.

5

Cantor's Transfinite Paradise

It was Cantor's work which gave sense to the question 'How many points are there in a line?', a question which previously lacked any precise sense. To avoid prejudging the question of whether Cantor should be seen as an inventor or a discoverer, the notion of 'sense' here can be treated as relating to the mathematician's understanding of the question. This was certainly changed by Cantor's work, whether we think that this was a change in the concepts involved or a more adequate grasping of unchanging concepts. But by looking at the way in which the question of the number of points on a line comes to have a sense we may be able to shed an indirect light on the discovery/invention issue as well as on the question of exactly what sense the question has been given.

Before Cantor developed his theory of transfinite numbers, the natural, and the only available answer, to the question was 'Infinitely many', and this was a way of saying that there is no number of points in a line, they are without number. This answer is not devoid of content for it indicates that given any finite number of points in a line there will always be more. In other words, the indeterminacy surrounding the totality of points in a line is of a specific kind and is unlike that surrounding the totality of angels that can dance on the head of a pin. One can at least start counting the points in a line, but cannot see any way to stop. So one knows that certain numerical answers are wrong, and to the extent that such answers are ruled out, the question has some sense; it is not wholly meaningless.

Thus we see that the first prerequisite for giving any further sense to the question is to have some more determinate conception of the totality of points in a line. In addition the concept of number needs to be extended or modified in such a way that even collections which cannot be fully counted may none the less be supposed to contain a determinate number of elements – may be assigned a number. The first of these, as was seen in chapter 4, is fulfilled by the development of the theory of real numbers. The second, which concerns us here, is effected by analyses of the concept of number which link numbering to the establishment of one–one correspondences rather than specifically to counting (counting is just one way of setting up a one–one correspondence between natural numbers and members of the set being counted).

1 Sets and Cardinal Numbers

The basic idea behind set theoretic analyses of the notion of positive whole number (natural number) is that it is to sets, or to collections of things, that numbers are assigned. A flock of sheep is counted and the number reached is the number of sheep in the flock. But it is also possible to compare sets in respect of the number of things they contain without actually counting them. One may, for example, establish that there are just as many cups as saucers on a tray by checking to see that there are no saucerless cups and no saucers without cups on them, without actually counting to find out how many of either there are. One can be in a position to say that there must be the same number of each without being able to say what that number is by establishing that there is a one–one correspondence between the set of cups on the tray and the set of saucers on the tray. Counting can then be seen as establishing the existence of a one–one correspondence between a finite set of objects and a subset of the natural numbers. If this correspondence can be thought to exist independently of anyone setting it up, then any finite set must contain a determinate number of objects, whether it has been counted or not. For any two finite sets it is obvious that there can be a one–one correspondence between them only if they contain the same number of elements, and it is tempting to generalize this to all sets, whether finite or infinite. This is precisely the step which Cantor took, but it is not entirely straightforward.

Suppose that we were to define 'number' by stipulating that a collection A has the same number of elements as another collection

B if (and only if) there is a one–one correspondence between them (a correlation which assigns to each member of *A* exactly one member of *B* and to 'each member of *B* exactly one member of *A*). Since this does not make any mention of whether the collections in question are finite or infinite it would seem to legitimize an extension of the notion of number into the infinite, enabling one to think of an infinite collection as containing a determinate number of things.

But the matter is not quite so simple. For as was mentioned in chapter 3, infinite sets are characterized by the seemingly paradoxical property of being such that they can be put into one–one correspondence with proper parts of themselves. This is paradoxical because it seems intuitively obvious that there must always be more elements in any whole than in some proper part of it.

The set of natural numbers $N = \{0, 1, 2, 3, \ldots n \ldots\}$ can be put in one–one correspondence with the set which consists only of the even numbers $E = \{0, 2, 4, 6, \ldots 2n, \ldots\}$ by pairing each n in N with $2n$, in E. Because E is infinite its supply of elements will never run out even though one would instinctively want to say that there must be twice as many elements in N as in E. And Dedekind defined infinite sets by reference to this characteristic: a set A is *infinite* if, and only if, there is a one–one correspondence between A and a set X which is a proper subset of A.

This means that if one were to say that two infinite sets contain the same number of elements when there is a one–one correspondence between them, and if one remains convinced that the size of any whole must always be greater than that of any of its proper parts, then the number of elements in an infinite set cannot be thought to be a measure of its size. For example, N and E will have the same 'number' of elements even though there are infinitely many numbers in N which are not in E, so that in this sense N is 'bigger than' E. This suggests that the elements of an infinite set are without number not just because they cannot be exhaustively counted but also because the notion of number, as a measure of size, can get no grip here. All infinite sets seem to come out as being of the same 'size' if one–one correspondence is taken as indicating sameness of size for sets. Indeed, if all infinite sets could be put into one–one correspondence with each other, one would be justified in treating the classification 'infinite' as an undifferentiated refusal of numerability. But given Cantor's discovery that there are infinite sets which cannot be put into one–one correspondence with each other, this conclusion is less compelling.

His proof, that for any set A (whether finite or infinite) there can be no one–one correspondence between A and the set of all subsets of A (the power set of A, denoted by $P(A)$), is important because it entails that there can be no one–one correspondence between the natural numbers and the real numbers. It immediately follows that the set N of natural numbers cannot be put into one–one correspondence with its power set $P(N)$. Since (a) each subset of the natural numbers can be uniquely correlated with an infinite sequence of zeros and ones, (b) each such sequence can be read as a binary decimal representation of a real number in the interval $(0, 1)$ and thus as representing a point on the unit line, and (c) the real numbers in $(0, 1)$ index all the points on whatever is chosen as the unit line, this means that the points on a line cannot be put into one–one correspondence with the natural numbers.

Cantor interpreted this impossibility of one–one correspondence as meaning that there must be 'more' points on a line than there are natural numbers (since there are clearly at least as many points in a line as there are natural numbers). More generally he interpreted it as licensing an attempt to extend the notion of number into the infinite. On this basis it became necessary to recognize a division between those sets which are denumerable, i.e. which can be put in one–one correspondence with the natural numbers, and those which, like the set of points in a line, are non-denumerable, i.e. for which no such correspondence exists.

Following Cantor in taking the first steps in this direction slightly more formally:

Definition   A *set*, or *aggregate*, is any collection into a whole M of definite and separate objects m of our intuition or thought.

Assumption   Every set, or aggregate, has a determinate 'power' or 'cardinal number'.

Definition   Two sets M and N have the same power, or cardinal number $(C(N) = C(M))$ if, and only if, there is a one–one correspondence between them.

Cantor's way of introducing this definition is to say:

> We will call by the name 'power' or 'cardinal number' of M the general concept which, by means of our active faculty of thought, arises from the aggregate M when we make abstraction from the nature of its various elements m and of the order in which they are given. (Cantor, 1955, p. 86)

If this is interpreted as an attempt to define a concept by reference to a mental act, something performed privately by each individual for himself, it is hardly a rigorous or adequate definition. But it can also be treated as a commentary on the precise condition under which two sets are to be said to have the same power. Whatever 'power' or 'cardinal number' is, it is a property of a set which does not depend on the specific nature of the elements it contains nor on the order in which the elements are given because neither of these are relevant to determining whether two sets are the same in this respect.

Likewise, the 'definition' of 'set' is less a definition than an attempt at explication of something which is being given the status of a primitive, undefined, term. For example Hausdorff introduces the term 'set' as follows:

> A set is formed by the grouping together of single objects into a whole. A set is a plurality thought of as a unit. (1957, p. 11)

What is implicit in both these explications is the thought that a set is a determinate collection of objects (a whole given after its parts) whose identity is entirely dependent on its members (the objects collected) and not on any method by which they may have been grouped or collected. Cantor, in particular, wanted to treat all sets as far as possible by analogy with finite sets. A finite set can be specified simply by listing its members, in the form $\{a, b, c, d\}$ where a, b, c, d need have nothing in common (a wasp, a London bus, Mount Everest, Mrs Thatcher). An infinite set might, by analogy, be thought to be specifiable by an infinite list. It might be just a contingent human limitation not to be able to think of an infinite set as a unit without going via some common characteristic of its elements, or some principle for selecting or generating them.

Here then, although classes, treated extensionally, would be counted as sets, sets are not restricted to being classes, i.e. are not restricted to being collections of objects which are the extensions of terms.

It is important to the development of set theoretic analyses of the natural numbers that sets be determinate collections of objects. Sets are to be just the sort of collections whose members can, or could in principle, be counted and thus assigned a number. But it is not easy to make the conditions of numerability explicit. To say that two sets have the same power if, and only if, there is a one–one correspondence between them does not yet entitle one to call these powers 'cardinal numbers' (where the sense of 'cardinal number' is derived from finite sets and natural numbers – the sense in which we may think of such sets as containing a determinate (finite) number of objects even though they have not been counted). At the very least we need to be able to say when the power, or cardinal number, of one set is greater or less than that of another.

Definition   Given any two sets A and B

$C(A) \leqslant C(B)$ iff there is a subset $B°$ of B such that
$C(B°) = C(A)$. *i.e.*
$C(A) \leqslant C(B)$ iff $(\exists B°)(B° \subseteq B \ \& \ C(B°) = C(A))$.
$C(A) < C(B)$ iff $C(A) \leqslant C(B)$ and $C(A) \neq C(B)$.

But to know that this defines even a partial order relation it is necessary to know that

$$C(A) \leqslant C(B) \ \& \ C(B) \leqslant C(A) \Rightarrow C(A) = C(B)$$

i.e. that if there is a subset $B°$ of B such that there is a one–one correspondence between A and $B°$ and there is a subset $A°$ of A such that there is a one–one correspondence between B and $A°$, then there is a one–one correspondence between A and B. For finite sets this is obvious, but not for infinite sets. The proof that it holds for infinite sets is known as the Schröder–Bernstein theorem (for a proof see, for example, Rotman and Kneebone, 1966, p. 49). But even this result, does not give all that is necessary for powers to look like cardinal numbers. Given any two sets A and B there are four possibilities:

1 There is $B° \subseteq B$ such that $C(A) = C(B°)$, and there is $A° \subseteq A$ such that $C(A°) = C(B)$.
2 There is $B° \subseteq B$ such that $C(A) = C(B°)$, but there is no $A° \subseteq A$ such that $C(A°) = C(B)$.
3 There is no $B° \subseteq B$ such that $C(A) = C(B°)$, but there is $A° \subseteq A$ such that $C(A°) = C(B)$.
4 There is no $B° \subseteq B$ such that $C(A) = C(B°)$, and there is no $A° \subseteq A$ such that $C(A°) = C(B)$.

We then have:

1 implies $C(A) = C(B)$
2 implies $C(A) < C(B)$
3 implies $C(B) < C(A)$
4 implies A and B are incomparable in respect of cardinality.

It can readily be shown that if either A or B or both are finite then case 4 will not arise, but the proof that 4 can never occur when both A and B are infinite requires a further assumption about sets – the assumption that every set can be well-ordered (ordered in such a way that each of its non-empty subsets has a least element).

Unable to prove the comparability of all sets in respect of cardinality, Cantor adopted it as an assumption. With this assumption it is possible to operate with powers in such a way that they do indeed begin to behave like cardinal numbers.

Definitions   Let A and B be sets. Let $C(A) = a$, $C(B) = b$, then

1 If A and B are disjoint, $a + b = C(A \cup B)$
  where $A \cup B = \{x : x \in A \text{ or } x \in B\}$.
2 $a \cdot b = C(A \times B)$
  where $A \times B = \{\langle x, y \rangle : x \in A \text{ \& } y \in B\}$.
3 $a^b = C(A^B)$
  where $A^B = \{f : f \text{ is a function from } B \text{ to } A\}$.

It can then be proved that powers, or cardinalities, behave very much as numbers should, and that in the case of finite sets we get all the results we should expect. If $A = \{a, b\}$ and $B = \{k, m, n\}$ and we put $C(A) = 2$, $C(B) = 3$ we find that

$$2 + 3 = C(\{a, b, k, m, n\}) = 5$$
$$2 \cdot 3 = C(\{\langle a, k\rangle, \langle a, m\rangle, \langle a, n\rangle, \langle b, k\rangle, \langle b, m\rangle, \langle b, n\rangle\}) = 6$$
$$2^3 = C(\{\{\langle k, a\rangle, \langle m, a\rangle, \langle n, a\rangle\}, \{\langle k, b\rangle, \langle m, b\rangle, \langle n, b\rangle\},$$
$$\{\langle k, a\rangle, \langle m, a\rangle, \langle n, b\rangle\}, \{\langle k, b\rangle, \langle m, b\rangle, \langle n, a\rangle\},$$
$$\{\langle k, a\rangle, \langle m, b\rangle, \langle n, b\rangle\}, \{\langle k, b\rangle, \langle m, a\rangle, \langle n, a\rangle\},$$
$$\{\langle k, a\rangle, \langle m, b\rangle, \langle n, a\rangle\}, \{\langle k, b\rangle, \langle m, a\rangle, \langle n, b\rangle\}\}) = 8$$

But, as might be expected, given that infinite sets can be put in one–one correspondence with proper subsets of themselves, infinite cardinalities do not behave quite like finite ones and their 'arithmetic' may seem a bit surprising. For example, if $\aleph_0 = C(N)$ where N is the set of natural numbers, then for any finite n

$$n\aleph_0 = \aleph_0 + \aleph_0 + \ldots (n \text{ times}) = \aleph_0$$
$$\aleph_{0^n} = \aleph_0 \cdot \aleph_0 \cdot \ldots (n \text{ times } n) = \aleph_0$$

Both the even numbers, E, and the odd numbers, O, can be put in one–one correspondence with the whole natural number sequence, i.e. $C(E) = C(O) = C(N) = \aleph_0$. And since $N = E \cup O$ and E and O are disjoint, $C(E) + C(O) = C(N)$, i.e. $\aleph_0 + \aleph_0 = \aleph_0$.

What this arithmetic of cardinal numbers does give is a way of expressing the relationship between the cardinality of a given set A and its power set $P(A)$ (the set of all subsets of A). For there is a one–one correspondence between subsets of A and the set of all functions f from A to a two element set, such as $\{0, 1\}$, where each subset is considered as the set of those elements of A for which f takes the value 1.

so $C(P(A)) = C(2^A) = 2^{C(A)}$

Since it has already been established that the cardinal number of the points on a (unit) line is the same as that of the real numbers in the interval $(0, 1)$, and that this in turn is the same as the cardinal number of the power set of the natural numbers $P(N)$, we can now put a 'number' on the points in a (unit) line, namely $C(P(N)) = 2^{\aleph_0}$.

But if, by analogy with $2^3 = 8$, we ask for an 'evaluation' of $2^{\aleph_0}$ we find that we do not have the means of supplying an answer. In particular Cantor thought that there are no infinite cardinal numbers in between $\aleph_0$ and $2^{\aleph_0}$, i.e. that $2^{\aleph_0}$ is the next infinite cardinal number after $\aleph_0$, but was unable to prove it. This is what has become known as Cantor's continuum hypothesis.

The situation so far is that it has just been assumed (for want of the means of providing a proof), that given any two sets A and B, either $C(A) = C(B)$, or $C(A) < C(B)$ or $C(B) < C(A)$. In other words it has been assumed that cardinalities, or cardinal numbers, can be arranged in a single linear order. But just making that assumption does not tell us anything about the nature of the cardinal number 'sequence', about how to establish where any given cardinality lies in it, or even whether it is correct to talk about there being a *next* cardinal number after $\aleph_0$. Our assumption does not rule out the possibility that infinite cardinalities might, like the rational numbers, be densely ordered. If that were the case, there would always be another cardinal number between any two given cardinalities and given any cardinal number there would be no 'next' one.

Once we get into the domain of infinite cardinalities the only procedure we so far have for reliably generating higher cardinalities is exponentiation – repeatedly taking the cardinal number of the power set of a given set. So we can form a series of sets

$N, P(N), P(P(N)), P(P(P(N))) \ldots$ with cardinal numbers

$\aleph_0, 2^{\aleph_0}, 2^{\aleph_0}, 2^{2^{2^{\aleph_0}}} \ldots$

In the finite case, exponentiation does not take us from one cardinal number to the next, there are lots of numbers in between 2 and 2^3, and even more between $2^3 = 8$, and 2^8. But we also know that infinite cardinalities do not behave in the same way as finite ones, so the question of whether there are or are not any infinite numbers in between those in the above series is an open question. The whole problem is that the theory of cardinality on its own has so effectively severed the connection between number and measure of size that it gives rise to no numerical scale by reference to which infinite sets might be 'measured'. The idea of a scale is one which involves an order; in demanding a scale we are asking for an ordering of cardinalities. But the infinite cardinalities were introduced by disregarding the associations of number with counting and with ordering, so it is perhaps not surprising that this way of introducing transfinite numbers yields 'numbers' which, even if ordered in reality, as the assumption of comparability asserts, cannot be put in order or named in order by us.

2 Transfinite Ordinal Numbers

However, as was seen in chapter 4, it was not with cardinal notions that Cantor first started to extend the notion of number into the transfinite (Cantor, 1883). His initial extension was of the natural number *sequence* into the transfinite, using numbers as a measure of the number of times an operation has been repeated. He first introduced his transfinite ordinal numbers as numbers which are *generated* in a sequence and thus as an extension of the natural number sequence, which is generated in counting by the principle of adding one to the previous number. Thus his first principle of generation for ordinal numbers is as follows.

First Principle of Generation   The addition of a unit to a number which has alredy been formed.

Used on its own this principle just gives us the ordinary natural numbers, or numbers belonging to what Cantor calls the first number class.

First Number Class $(I) = 0, 1, 2, 3, 4, \ldots$

His second principle of generation is one which allows for the formation of the first infinite ordinal numbers as limit numbers. We imagine that the set of natural numbers can be run through in order and, assuming they constitute an actually infinite set, that there must be an infinite bound to the numbers required. The second principle of generation allows for the formation of a 'number', ω, to stand for the first number which is greater than all the finite numbers. ω is thought of as a limit which the sequence $0, 1, 2, 3, \ldots$ approaches but never attains.

Second Principle of Generation If there is defined any definite succession of real integers of which there is no greatest, a new number is created, which is defined as the next greatest to them all.

Once one of these new, infinite ordinal numbers has been introduced, the first principle of generation will apply to it so that we get a new sequence $\omega + 1$, $\omega + 2$, $\omega + 3$, ... This, being a succession of integers with no greatest member, is a sequence to which the second principle can be reapplied so that we get another infinite ordinal number $\omega + \omega$ or $\omega \cdot 2$. The infinite ordinal numbers generated by repeated applications of these first two principles alone form what Cantor called the second number class.

Second Number Class $(\text{II}) = \omega$, $\omega + 1$, ..., $\omega + n$, ..., $\omega \cdot 2$, $(\omega \cdot 2) + 1$, ... $\omega \cdot 3$, ..., $\omega \cdot \omega$, ...

All the numbers in the second number class can, however, be thought of as the numbers obtained by introducing a more or less complicated order on the sequence of natural numbers. Any sequence which, like 0, 1, 2, 3, ..., is a linear sequence with no last member and which involves only one infinite sequence will be called a *simply infinite sequence*. The sequence formed by running through all the natural numbers from 2 on and then tacking 0 and 1 on at the end after all the rest, is not a simply infinite sequence but one whose ordinal number is $\omega + 2$, i.e. a simply infinite sequence followed by a two element sequence (2, 3, 4 ... 0, 1). Similarly the sequence formed by running through all the even numbers and then all the odd numbers, (0, 2, 4, ... 1, 3, 5, ...), has the ordinal number $\omega + \omega$, i.e. it is one simply infinite sequence followed by another. The sequence formed by running through all the numbers divisible by 2, followed by all those divisible by 3, followed by all those divisible by 5, and so on for all the prime numbers.

$(2, 4, 6, \ldots 3, 6, 9, \ldots 5, 10, 15, \ldots \quad \ldots \quad \ldots)$

is a simply infinite sequence of simply infinite sequences, whose ordinal number is $\omega \cdot \omega$, since there is no greatest prime number. This means that although we have generated a lot of infinite ordinal numbers (numbers which depend on the order in which a set is given) they are all such that they are ordinal numbers of sets which can be put in one–one correspondence with the natural numbers (denumerable sets), and indeed are all ordinal numbers which can be assigned to the set of natural numbers when it is listed (or 'counted') in something other than its natural order.

So what we have is a proliferation of infinite ordinal numbers which all apply to sets having the same cardinality, $\aleph_0$. These first two principles on their own do not generate any ordinal number which could be the number of points in a line, since this set has a cardinality greater than $\aleph_0$. Thus Cantor introduces a third principle of generation, which he also called the principle of limitation, or the principle of interruption.

Third Principle – Principle of Limitation   All the numbers formed next after ω should be such that the aggregate of numbers preceding each one should have the same power (or cardinality) as the first number class. These numbers constitute the second number class.

The idea behind this principle is to delimit a totality of ordinal numbers produced by the first two principles (the second number class) in such a way that the second principle can then be applied to give a new number (ω_1) which is defined as the next number greater than all of the numbers belonging to the second number class. The first two principles can then be reapplied to further extend the ordinal sequence. A more general form of the principle of limitation allows this process to go on indefinitely by sectioning the numbers generated into number classes to which the second principle can then be applied.

General Principle of Limitation   All the numbers formed next after ω_α should be such that the aggregate of numbers preceding each one should have the same power (or cardinality) as the $(\alpha + 1)$th number class. These numbers then form the $(\alpha + 2)$th number class.

Cantor proved that the second number class cannot be put into one–one correspondence with the first and that there can be no set with a cardinality in between those of the two number classes. More generally he proved that the cardinality of the $(\alpha + 1)$th number class will be greater than that of the αth number class and that there can be no set with a cardinal number in between these. Thus the cardinal number of the second number class is the next after $\aleph_0$, and is labelled $\aleph_1$. Any set whose ordinal number is ω_1 or more will also have a cardinality greater than $\aleph_0$, i.e. will be a non-denumerable set.

3 Ordinal Numbers and Cantor's Continuum Hypothesis

Our question about the number of points in a line thus now receives a further sense. We can now ask whether its cardinality is the next after $\aleph_0$ by asking whether $2^{\aleph_0} = \aleph_1$. The more precise form of Cantor's continuum hypothesis asserts that this is the case. We can also ask whether it is possible to assign the points in a line, in their natural order, an ordinal number. If this were possible then it would be relatively easy to answer the question about the cardinality of the continuum because we would merely have to know which number class the relevant ordinal number falls into. The problem is that the points in a line in their natural order cannot be assigned an ordinal number, even one which involves going through infinitely many infinite sequences.

To see this it is necessary to examine more closely what is involved in assigning to a set an ordinal number, making explicit some of the assumptions which have been made. The ordinal number sequence was extended into the infinite by imitating as closely as possible the ordinary natural number sequence. Similarly the application of infinite ordinal numbers to sets is to imitate as closely as possible the application of the finite, natural numbers when they are used in counting to give the number of elements in a finite collection.

When the elements of a finite set are counted this can be seen as setting up a one–one correspondence between these elements and the first part of the natural number sequence given in order. When the elements of the set are exhausted the last number used gives the number of elements in the set. This process of counting involves selecting the elements counted in a particular order; when a set is counted it is counted in a particular order. When the set to be 'counted' is infinite the process of counting does not stop so one cannot assign a number on the basis of saying that it is the last one to be used. But one can continue to use the idea of putting the elements of the set in an order which matches the order of the first part of the ordinal number sequence up to some specific ordinal number. This was what was done with the examples of the different orderings (ways of counting) the natural number sequence, and is the idea used in extending the ordinary arithmetic operations to the ordinal numbers. This is based on the idea that the addition of ordinal numbers involves placing two ordered sequences end to end to obtain a new, extended ordered sequence. Thus if $A = \langle a, b \rangle$ and $B = \langle c, d, e \rangle$ are ordered sets, the ordered union $A + B = \langle a, b, c, d, e \rangle$, whereas the ordered union $B + A \langle c, d, e, a, b \rangle$. This means that the addition of ordinal numbers will not, in the infinite case, be commutative. Multiplication of ordinal numbers is defined as repeated addition. Where infinite sequences are involved the ordinary laws of arithmetic are not all obeyed. For example:

$$1 \cdot \omega = 1 + 1 + 1 \ldots = \omega$$
$$2 \cdot \omega = 2 + 2 + 2 + \ldots = (1 + 1) + (1 + 1) + (1 + 1) =$$
$$\ldots = \omega$$
$$\omega \cdot 2 = \omega + \omega = (1 + 1 + 1 + \ldots) + (1 + 1 + 1 + \ldots)$$
$$\omega \cdot (\omega + 1) = (\omega + \omega + \omega + \ldots) + \omega = (\omega \cdot \omega) + \omega$$
$$(\omega + 1) \cdot \omega = (\omega + 1) + (\omega + 1) + (\omega + 1) + \ldots = \omega \cdot \omega$$

For a set to be assigned an ordinal number it must be possible to order it in the same kind of way that the sequence of ordinal numbers is ordered. The ordinal number sequence is constructed by starting a sequence, letting it run on infinitely and then taking the next number after all of those and starting again. There are lots of bits of the sequence which have no greatest member (there is no greatest natural number, for example) but every bit has a least member (it always starts somewhere). So if a set is to have an ordinal number it must be possible to arrange its elements in a linear order which is such that every non-empty subset has a least element, i.e. to impose a well-ordering on it.

But the points in a line (or the real numbers in the interval (0, 1)) given in their natural order are not well-ordered. There are indefinitely many subsections of the line which have no first point. For example the set of points corresponding to numbers greater than $\frac{1}{2}$ and less than 1. Since there is no real number immediately after $\frac{1}{2}$ (between any given number and $\frac{1}{2}$ there will always be another number) there is no least element to this set. So if the points in a line were to be assigned an ordinal number it would have to be possible to impose on them an order which is different from their natural order and which is a well-ordering. This can be done for the rational numbers, for they too are not well-ordered by the natural order; there is no least rational number greater than $\frac{1}{2}$ and less than one. But it is possible to write out all positive rational numbers (with some repetitions) using two dimensions:

$$
\begin{array}{ccccccc}
 & 1 & 2 & 3 & 4 & 5 & 6 & \ldots \\
1 & 1/1 \rightarrow 1/2 & 1/3 \rightarrow 1/4 & 1/5 \rightarrow 1/6 & \ldots \\
2 & 2/1 & 2/2 & 2/3 & 2/4 & 2/5 & 2/6 & \ldots \\
3 & 3/1 & 3/2 & 3/3 & 3/4 & 3/5 & 3/6 & \ldots \\
4 & 4/1 & 4/2 & 4/3 & 4/4 & 4/5 & 4/6 & \ldots \\
5 & 5/1 & 5/2 & 5/3 & 5/4 & 5/5 & 5/6 & \ldots \\
6 & 6/1 & 6/2 & 6/3 & 6/4 & 6/5 & 6/6 & \ldots
\end{array}
$$

The numbers in the array can be listed by following the arrows, giving the sequence:

$$
\begin{array}{ccccccccccc}
1/1, & 1/2, & 2/1, & 3/1, & 2/2, & 1/3, & 1/4, & 2/3, & 3/2, & 4/1, & 5/1, \ldots \\
1 & 2 & 3 & 4 & 5 & 6 & 7 & 8 & 9 & 10 & 11 \ldots
\end{array}
$$

And this gives not just a well-ordering but also a one–one correspondence with the natural numbers in their natural order. Each rational number can be expressed in a form x/y, where x and y are relatively prime. The numbers of this form constitute an infinite subset of that listed (infinite because $n/1$ is included for each natural number n). The listing thus effects a one–one correspondence between the positive rational numbers and a subset of the natural numbers in their natural order. Thus the rational numbers, ordered in this way, have ordinal number ω and are shown to be denumerable, i.e. to have cardinality $\aleph_0$.

However, it is not possible to produce an ordering on the real numbers in the same sort of way, for if it were, there would be a one–one correspondence between them and the natural numbers.

One can use a square array to show that no enumeration of the real numbers in (0, 1) can be complete; there must always be a number which has been missed out. Suppose that the real numbers are given in terms of their binary decimal expansion, then when we list them we get an array:

	r_1	r_2	r_3	r_4	...
1	0	1	1	0	...
2	0	0	1	1	...
3	1	1	1	0	...
4	1	0	1	1	...

Define a new number r given by reading down the diagonal of the array and interchanging zeros and ones, i.e. given by the decimal expansion .1100.... This number differs from all those listed because, for each n, it differs in the nth decimal place from r_n. This means that there are no immediate grounds for supposing either that there is a well-ordering of the points in a line or that there is not.

The extension of the ordinal number sequence into the transfinite gives a way of generating not only infinite ordinal numbers, but also a sequence of infinite cardinal numbers, the numbers of the successive number classes, which is such that each one is the 'next greatest' after the one which it follows. So the ordinal number sequence also provides a scale of cardinal numbers on which one might hope to locate $2^{\aleph_0}$. But the problem, as Cantor saw, was that the two routes to cardinal numbers are largely independent. There is no guarantee that every cardinality (the power of every set) will have a representative amongst the cardinalities of the number classes (classes of ordinal numbers) generated by Cantor's three principles. Without such a guarantee one has no reason to suppose that there must be a definite answer to 'Where on the sequence of alephs indexing the number classes, does $2^{\aleph_0}$ lie?' In proposing that $2^{\aleph_0} = \aleph_1$ Cantor also assumed that every set, not just the set of points on a line, can be well-ordered, and it was this assumption that entitled him to claim that every set, including the set of points on a line, has at least one ordinal number. Any set which has an ordinal number will have a cardinal number which is amongst the alephs. If it is not the case that every set can be well-ordered, then the cardinalities of those sets which cannot be so ordered will not be represented by cardinal numbers arising from the construction of the ordinal number sequence.

But if the ordinal number sequence is just a construct, something Cantor brought into existence by defining certain symbols, principles for generating more of them and rules for manipulating them, there would be an essential asymmetry between the status of infinite cardinal and infinite ordinal numbers. This asymmetry would not justify Cantor's persistent attempts to prove his continuum hypothesis. For he believed not only that it was true but that, as a mathematical truth, it should also be provable.

4 Order Types

Cardinal numbers were introduced in terms of one–one correspondences between independently given sets. The continuum hypothesis seeks to equate the cardinality of one such independently given set with that of a set of ordinal numbers, a constructed set. But if the ordinal numbers are purely mental constructs there are no grounds for supposing that this equation can be *proved* as a mathematical theorem, any more than one would suppose that the number of planets can be mathematically proved to be nine. Only those properties of ordinal numbers which follow from the way in which they have been constructed could be expected to be provable, not anything about their relation to independently given sets. This raises the philosophically crucial question of whether it makes sense to suppose that there are mathematical truths which are not provable. Could the continuum hypothesis be true (or false) but not provably one or the other? If it were true but not provable (Cantor believed it to be true but could not prove it) what sort of grounds, if any, could we have for thinking it to be true? These are questions whose discussion is to be postponed until chapter 9, but which can never be very far away when exploring a newly created branch of mathematics (the study of transfinite numbers, the discipline of transfinite set theory, was created by Cantor and his successors, whether the object of their study was created by them or not).

But although Cantor first introduced the transfinite ordinals via principles of construction, he did not regard these numbers as *mere* products of mental construction. Their justification to the title 'number' requires more than generation in a sequence and more than definition of 'arithmetic' operations as purely symbolic manipulations. They must function something like numbers in that they can be applied and seen as giving a certain kind of information about the sets to which they are applied. In other words they have to be shown to have a use. This requires that the arithmetic operations be interpretable as operations applied to well-ordered sets. To make the parallel with cardinal numbers closer Cantor gives a definition, in terms of one–one correspondence, of what it is for two sets to have the same ordinal number.

Definition   Two well-ordered sets A, ordered by $\leqslant_1$, and B ordered by $\leqslant_2$, have the same order type (or ordinal number) ($O(A, \leqslant_1) = O(B, \leqslant_2)$) if, and only if, there is a one–one, order preserving correspondence c between A and B, i.e. c is such that for every x, y in A, $x \leqslant_1 y$ iff $c(x) \leqslant_2 c(y)$.

Definition   The segment A_a of an element a of a well-ordered set $\langle A, \leqslant \rangle$ is the subset of A which consists of those elements of A which precede a, i.e.

$$A_a = \{x : x \in A \ \& \ x < a\}$$

It can then be proved that a well-ordered set does not have the same order type as any of its segments, although in the case of an infinite set it is still possible for the whole set to have the same order type as one of its proper subsets (one which is not a segment). For example, take the set N of natural numbers given in their natural order. Any segment of N will be a finite set, the set of numbers less than n for some natural number n. Since every finite set has a greatest member and N does not, none of these segments is of the same order type as N. But the set of even numbers does have the same order type as N, since the mapping $f(n) = 2n$ from N to E is one–one and order preserving.

Definition   If $\langle A, \leqslant_1 \rangle$ and $\langle B, \leqslant_2 \rangle$ are well-ordered sets then $O(A, \leqslant_1) < O(B, \leqslant_2)$ if, and only if, A has the same order type as some segment of B, i.e. if, and only if, there is an element b of B, such that $O(A, \leqslant_1) = O(B_b, \leqslant_2)$

The comparability of well-ordered sets can then be proved, i.e. given any two well-ordered sets $\langle A, \leqslant_1 \rangle$ and $\langle B, \leqslant_2 \rangle$ either

$$O(A, \leqslant_1) = O(B, \leqslant_2), \text{ or } O(A, \leqslant_1) < O(B, \leqslant_2), \text{ or}$$
$$O(B, \leqslant_2) < O(A, \leqslant_2).$$

The ordinal numbers are themselves well-ordered and, in the finite case, the set of numbers which are less than n, $\{0, 1, 2, \ldots n - 1\}$, itself contains n members. This suggests using the ordinal number sequence as a standard well-ordered set, a scale against which all others can be compared, and thereby assigned an ordinal number.

Definition A set M, well-ordered by $\leqslant$, has ordinal number α if, and only if, M ordered by $\leqslant$ has the same order type as the set of ordinal numbers less than α under their natural ordering.

Addition and multiplication of ordinal numbers can now be associated with operations on well-ordered sets.

Definition If $O(A, \leqslant_1) = \alpha$ and $O(B, \leqslant_2) = \beta$ and A and B are disjoint then $\alpha + \beta = O(\langle A \cup B \rangle)$, where $\langle A \cup B \rangle$ is the ordered union of A and B, i.e. $A \cup B$ ordered by the relation $\leqslant$ defined as follows: $x \leqslant y$ if, and only if (a) $x, y \in A$ & $x \leqslant_1 y$, or (b) $x, y \in B$ & $x \leqslant_2 y$, or (c) $x \in A$ & $y \in B$.

In other words $\alpha + \beta$ is the ordinal number of the well-ordered set which results from first running through all of A and following this by all of B in their given orders, and this definition holds whether A and B are sets of ordinal numbers or of objects of other kinds. Multiplication was defined in terms of repeated addition: $\alpha \cdot \beta = \alpha + \alpha + \alpha \ldots \beta$ times. Thus $\alpha \cdot \beta$ will be the ordinal number of the ordered union of a whole sequence, whose ordinal number is β, of disjoint sets, each with ordinal number α.

It is now also possible to give an alternative definition of cardinal number.

Definition   The cardinal number $C(X)$ of a set X is the least ordinal number α such that there is a one–one correspondence between X and $\{x : x$ is an ordinal number and $x < \alpha\}$.

This definition can replace the preceding one, on the assumption that every set can be well-ordered. If this assumption fails then either it has to be allowed that there are some sets which lack cardinal numbers or it has to be allowed that there are cardinal numbers which cannot be identified with ordinal numbers.

In this way the theory of infinite ordinal and cardinal numbers can be integrated, but this is done in such a way that the basis for further investigation lies in a study of sets. For questions about what numbers there are, and what their relations are, have been made dependent on questions concerning what sets exist and what are the relations between them. Even the generation of the ordinal number sequence is dependent on the power of the principle of limitation to mark off the ordinal numbers into sets, or classes, to which the second principle of generation can then be applied. So the question of what ordinal numbers there are depends, for its answer, on an answer to the question 'What classes of ordinal numbers are there?'

5 Set Theoretic Paradoxes

Moreover, this becomes a pressing question in the light of what has become known as the Burali–Forti paradox. If one supposes that there are no limitations on the formation of classes of ordinal numbers, then it must be the case that there is a class consisting of all the ordinal numbers. But if these numbers can be 'limited' to form a class, it is certainly a well-ordered class and so must itself have an ordinal number. The second principle of generation would assign it a new number, that which is the next greatest after all the ordinals But there can be no ordinal number greater than all the ordinals. Yet the ordinal number, Ω, of the set of all ordinals cannot be an ordinal number belonging to that set, for then it would be a well-ordered set which has the same ordinal number as a segment of itself, $\{x : x$ is an ordinal number & $x < \Omega\}$. But it can be proved that no well-ordered set can have the same ordinal number as a proper segment of itself. So if the theory of ordinal numbers is to be consistent, clearly the totality of ordinal numbers cannot be allowed to form a set or class to which the second principle of generation can be applied or to which the notion of ordinal number can be applied. Some more precise delimitation of the permissible 'limitations' of the ordinal number sequence is required. The second principle of generation is what allows entry into the transfinite domain; without it there would be no infinite ordinal numbers. But the vagueness inherent in it concerns what is to count as a defined definite succession of real integers; clearly not every candidate for being a definite succession of ordinal numbers will do.

The problem is not limited to the ordinal numbers, however. A very similar situation, exhibited by Cantor's paradox, occurs in the case of cardinal numbers. Cantor proved that for any set A, the cardinal number of $P(A)$ is strictly greater than that of A, $C(P(A)) > C(A)$. This entails that there can be no greatest cardinal number, for given a set of no matter what cardinality, its power set will have a greater cardinality. But consider now the set U consisting of all objects (including classes). This is the greatest possible set, since it includes all other sets as subsets. Moreover for any two sets A and B, if A is a subset of B, $C(A) \leqslant C(B)$. So that every set must have a cardinal number less than or equal to that of U. Hence $C(U)$ is the greatest cardinal number, contradicting our previous conclusion. So again consistency would seem to require that the set U should not be treated as a totality to which the notion of number can be applied.

These paradoxes clearly pose a threat to the whole theory of transfinite numbers. At the very least their claim to mathematical legitimacy requires that use of these notions should not lead to contradictions. Even the most ardent proponent of the view that mathematics is a free creative activity recognizes the consistency constraint – the mathematician can create what realms he wants provided they are free from contradiction. In a pure dream world contradictions don't matter, but if mathematicians are to go in for giving proofs, and if their theories are to be used, then contradictions must be avoided. However, the appearance of contradictions in a new theory does not condemn it at once. It is quite possible that it has been incorrectly formulated, or that some inappropriate assumption has been made. In any new development, whether of a game, a piece of legislation, a computer program or a radar system, the prototype will need improving, modifying and generally tidying up before it will function properly.

Cantor's reaction to the paradoxes was to introduce a distinction between the ordinary infinite, which is, in his view, a proper domain of mathematical study and subject to the numerical methods proposed in his transfinite arithmetic, and the absolute infinite which is beyond all numbering, measuring and human reasoning. He thus effectively extends the notion of number so that it can now be applied to totalities which previously had to be regarded as being without number, but even so there still have to be some totalities which are without number. The members of any absolutely infinite collection are without either ordinal or cardinal number. But this still leaves a problem which is how to tell whether a given collection is absolutely infinite or just ordinarily infinite.

That the fundamental difficulty here lies not so much with the notion of number as with the notion of set to which it has been inseparably linked in Cantor's theory, is suggested by Russell's paradox which does not involve numbers at all, only sets. This paradox arises by modifying the proof (given on pp. 63–4) that there can be no one–one correspondence between a set A and its power set $P(A)$. Since sets (power sets in particular) can contain other sets as members, it seems sensible to ask whether a given set belongs to itself. For example the set of all sets containing more than two elements must belong to itself, whereas the set of sets containing less than two elements does not. Consider then the set R of all those sets which do not belong to themselves. R must either belong to itself or not. If R belongs to itself, then it satisfies its own defining conditions, i.e. it does not belong to itself. So R cannot belong to itself, but then it does satisfy its own defining condition and does belong to itself. So we are caught in a contradiction either way. A consistent set theory cannot then allow R to exist, for it is a contradictory set. But this raises the difficult question of the grounds on which R is to be excluded. It will not be enough simply to legislate R out of existence since there might well be other problematic sets waiting to surprise us. To be sure of getting a consistent set theory, something without which the theory of transfinite numbers can never be assured of consistency, there have to be general principles governing set existence. These principles have either to be laid down or discovered by appeal to some more general considerations.

There are two slightly divergent concerns here. There is the concern of the mathematician to get a working theory, one about which he can be reasonably confident, and there is that of the philosopher worried about the nature and status of this mathematical activity. He wants to know not merely what principles will produce a workable set theory, but what sort of justification, if any, these principles can have, whether they are ultimate first principles or whether they can be justified by appeal to some more basic principles. A justification from principles known to be reliable would constitute a guarantee of consistency, but in the absence of any justification he will want to know what possible assurance there can be of consistency.

The mathematical approach to systematizing and rigorizing a body of unsystematized procedures is that established with Euclid, namely, axiomatization. Producing a minimal list of axioms or postulates about what exists and about what operations can be performed and then proceeding systematically to show that most of the previously used results, together with many others, can now be proved from these axioms. It is this approach which gives the question concerning the number of points on a line its present meaning and so this is what will be sketched in the next chapter. The more philosophical concerns, which have been raised here will be postponed until we have a clearer view of the mathematical situation.

6

Axiomatic Set Theory

The axiomatic approach to set theory was initiated by Zermelo. A number of variant systems of axioms have been proposed but that which has become standard is basically Zermelo's original system with modifications introduced by Fraenkel and which is thus known as Zermelo–Fraenkel set theory, or ZF for short.

1 Axiomatization

Any set of axioms may be viewed in more than one way. They might be seen (a) as expressing basic truths about a universe of objects of a certain kind which exist independently of the mathematician's thought or of his constructions, or (b) as giving the basic building blocks and principles for constructing a universe of objects of a certain kind. They might even be seen (c) as giving the rules of a 'game' to be played with newly introduced symbols; rules which determine the permissible symbolic moves in the game of constructing proofs. But this last view would give us no reason for wanting to play the new game or for thinking that this should be taken seriously to be part of mathematics as a discipline studied with the aim of acquiring knowledge of a kind which is usually distinguished from the sort of skill possessed by a chess master. Formalists, however, have argued that this distinction is one which cannot be sustained, as indeed it could not be if it could be generally shown that there can be no such thing as mathematical knowledge, or that it is misleading to talk of knowledge here because it suggests that mathematics is similar to the natural sciences and contains something analogous to factual knowledge.

To pursue this matter thoroughly would require an investigation into what was meant by knowledge in the natural sciences as well as in mathematics and also into the role of symbols in both the expression and acquisition of knowledge and skills of different kinds. This is something which is well beyond the scope of the present discussion. Moreover, in the context of discussions of set theory, formalism represents something of a side-track in that the predominant attitude adopted has been to assume that axiomatization here is not a matter of starting from scratch to create something completely new, but is rather a matter of trying to tidy up, by formalizing and systematizing, already existing concepts and the practices within which they are embedded. To this extent the axiomatization of set theory is not unlike the axiomatization of geometry.

Euclid's axiomatization of geometry can either be seen as an attempt to enunciate the fundamental truths about space, principles which will characterize it uniquely and so allow for the proof or disproof of any proposed geometrical proposition, or they may be seen as giving the basic principles of a theory of geometrical contruction: postulating two basic operations, the construction of straight lines and circles, by ruler and compass, and giving the principles necessary to show how further, more complex figures can be constructed (and thus proved to exist) and their properties investigated. The formalist would claim that they serve to introduce definitions of the basic geometrical terms, 'point' and 'line', and give principles for using them. But in Euclid's case there would be a clear problem about the adequacy of the definitions. For definitions, even when set out in the form of axioms, are given in words, whereas from the outset Euclid's proofs involve not just words but also diagrams.

An essential step in many proofs is a geometrical construction introducing extra lines or circles to help prove the correctness of a claim about the properties of a given figure. Rigorously adequate definitions of new symbols which are intended to do more than merely allow us to play a game of symbol manipulation, definitions which fix their use, should, it would seem, be able somehow to tie the symbols to the domain we are interested in so that we can actually use them to clarify thinking in a given area. Euclid's words need to be attached to his diagrams and to operations for constructing diagrams. This attachment can be made if it is assumed that the undefined terms, 'point' and 'line', have an antecedent and under-stood meaning which is being refined by the axioms but not being fully given by them. Thus in the case of Euclid's axiomatization of geometry it is really only possible to adopt attitudes (a) or (b) because the axiomatization is clearly introduced to bring order and increased clarity and rigour to an area of thought which already exists.

Set theory was in a very similar position. Cantor and others were already using notions of set, class or aggregate assuming that they were already sufficiently clear and sufficiently precisely defined. But the new and extended uses to which they were putting these terms revealed imprecisions and difficulties in the form of paradoxes. More clearly delimited principles of use were required but ones which would be constrained by accepted uses. It was a question of reforming existing terminology and its associated practices, rather than introducing a completely new system of terms.

The need for set theory itself, however, emerged only as a result of the move fully to formalize geometry, analytic geometry in particular, and to render it independent of intuition. A fully formal axiomatization of geometry (one in which proofs could all, in principle, be produced as deductions written within a formal language, whose primitive terms were implicitly defined by the axioms determining their use, without need to appeal to con-structions, figures or diagrams) was produced by Hilbert in 1899. This was made possible only by reference to the theory of point sets and the kind of analysis of the structure of relations between the points of a continuum on which Cantor was engaged. The role of set theory as a vital part of the programme to eliminate from mathematics appeals to intuition (in particular to geometrical intuition) raises difficult questions about the ground on which intuitions about sets (those intuitions which are appealed to in the selection and adoption of axioms) can be supposed to be superior to, or mathematically more reliable than, geometrical intuition.

These questions could be circumvented by taking a wholly formalist attitude toward set theory itself, but this is not a possibility prior to an axiomatic formalization of the theory. Intuition (merely as an informal grasp of concepts, whatever that might consist in) has, at the very least, to be the ladder climbed in reaching an axiomatization, even if it is to be dispensed with later. The difference between producing a set theory which provides a useful framework for working mathematicians and one which is a plausible candidate for a formally consistent system – one in which paradoxes do not obviously arise – is illustrated by the contrast between the Zermelo–Fraenkel axiomatization and the system called New Foundations (NF) produced by Quine, which uses formal devices to restrict the formation of problem sets. Quine's system has never been used by mathematicians, who even had difficulty investigating its properties and comparing it with ZF. The intuitions underlying the adoption of the ZF axioms will be discussed more fully in chapters 7 and 8.

2 The ZF Axioms

In presenting the ZF axioms it is presumed that the domain of entities of which they are true is a universe of sets. All individual variables $x, y, \ldots$ are thus interpreted as ranging over the universe of sets. There is a single undefined relation '$\in$' which is intended to stand for the membership relation, so that '$a \in b$' means 'a is an element of b' or 'a belongs to b'. The relation of set inclusion '$\subseteq$' or the subset relation can be defined in terms of '$\in$'. If two sets b and c are such that every element of b is also an element of c, then b is said to be included in, or to be a subset of, c. (Formally $b \subseteq c \equiv_{df} \forall x (x \in b \Rightarrow x \in c)$.)

1   *Axiom of extensionality*   If two sets have the same elements then they are identical.

$$\forall x \forall y \forall z [(z \in x \Leftrightarrow z \in y) \Rightarrow x = y]$$

2   *Null set axiom*   There is an empty set, one which contains no elements.

$$\exists x \forall y \neg (y \in x)$$

It can be shown that there can only be one such set and so it will subsequently be denoted by $\emptyset$.

3   *Pair set axiom*   If a and b are sets then there is a set $\{a\}$ whose only element is a and there is a set $\{a, b\}$ whose only elements are a and b.

$$\forall x \forall y \exists z \forall w (w \in z \Leftrightarrow w = x \lor w = y)$$

4   *Sum set axiom*   If a is a set then there is a set $\cup a$, the *union* of all the elements of a, whose elements are all the elements of elements of a.

$$\forall x \exists y \forall z [z \in y \Leftrightarrow \exists w (w \in a \ \& \ z \in w)]$$

5   *Axiom of infinity*   There is a set which has $\varnothing$ as an element and which is such that if a is an element of it then $\cup \{a, \{a\}\}$ (or $a \cup \{a\}$) is also an element of it.

$$\exists x [\varnothing \in x \ \&$$
$$\forall y (y \in x \Rightarrow \exists z (z \in x \ \& \ \forall w (w \in z \Leftrightarrow w \in y \lor w = y)))]$$

6   *Axiom of foundation*   If a is a non-empty set, then there is an element b of a such that there are no sets which belong both to a and b.

$$\forall x [\neg (x = \varnothing) \Rightarrow \exists y (y \in x \ \& \ \forall z (z \in x \Rightarrow \neg (z \in y)))]$$

7(a) *Subset axiom*   If a is a set and $F(x)$ is any well-formed expression in the language of ZF with a single free variable, then there is a set b whose elements are those elements of a for which $F(a)$ is true.

$$\forall x \exists y \forall z [z \in y \Leftrightarrow z \in x \ \& \ F(z)]$$

7(b) *Replacement axiom*   If a is a set and $F(x, y)$ is a well formed expression in the language of ZF which associates with every element x of a a unique element $x°$, then there is a set $a°$ whose elements are just those sets $x°$ associated by $F(x, y)$ with elements of a.

$$\forall x [\forall y [y \in x \Rightarrow \exists z (F(y, z) \ \& \ \forall w (F(y, w) \Rightarrow w = z))] \Rightarrow$$
$$\exists v \forall u [u \in v \Leftrightarrow \exists t (t \in x \ \& \ F(t, u))]]$$

8   *Power set axiom*   If a is a set, then there is a set $P(a)$, the *power set* of a, whose elements are all the subsets of a.

$$\forall x \exists y \forall z [z \in y \Leftrightarrow \forall w (w \in z \Rightarrow w \in x)]$$

9   *Axiom of choice*   If a is a set, all of whose elements are non-empty sets no two of which have any elements in common, then there is a set c which has precisely one element in common with each element of a.

$$\forall x [\forall y (y \in x \Rightarrow \neg (y = \varnothing)) \& \forall y \forall z (y \in x \& z \in x \, (y = z) \Rightarrow$$
$$\neg (\exists w (w \in y \& w \in z))) \Rightarrow \exists u \forall y (y \in x \Rightarrow \exists z (z \in u \&$$
$$z \in y \& \forall w (w \in u \& w \in y \Rightarrow w = z)))]$$

The concern underlying the ZF axiomatization was to delimit the domain of sets with which mathematicians were already concerned or might conceivably be concerned in such a way as to be sure of excluding the problem, paradox producing, cases. Whether this is to be thought of as a matter of describing a set theoretic universe, reaching a correct concept of set, or of constructing a universe in which the mathematician can work then depends crucially on the attitude taken to mathematics as a whole. If mathematics is thought, in general, to be a matter of discovering and correctly describing what exists already, then, if mathematicians find it necessary to use the concept of set, sets must be thought to be part of this pre-existing mathematical reality. If mathematics is more a matter of creation and construction, then the development of a more adequate set theory will be a part of this ongoing process.

The difference between these two attitudes to the axiomatization of set theory is not very great when it comes to understanding how the axioms work and why they have been selected. The axioms either straightforwardly assert the existence of a set satisfying a particular condition (null set and infinity) or say that given any set there is another one related to it in a specified way. Looking at this in terms of construction, we are given some basic building blocks together with ways of constructing new sets out of ones already constructed. Looking at it in terms of describing a universe of sets the axioms assure the mathematician that certain sets do exist and that given any other set, whose existence he can prove, sets related to it in specified ways also exist. From both points of view the mathematician is being given a framework within which it is clear how to prove the existence of a set.

The real difference comes in determining what critical standards to apply to an axiomatization and in the attitude to be taken with regard to the adoption or non-adoption of the more controversial axioms. From the constructive point of view the only sets which exist are those which can be shown to exist by repeated application of the already adopted axioms. The question regarding the adoption of further axioms may be complicated, but is basically a question of what will be useful or of what is required by other areas of mathematics. From the non-constructive point of view there may be many other sets which exist but whose existence cannot be proved from an existing list of axioms so that a question remains about what sets do exist and of how this can be known; how is it to be determined what further axioms should be adopted and how could it be known when the list is complete?

The *axiom of extensionality* is designed to clarify what a set is – it is a collection of elements whose identity is wholly determined by those elements. This means that the way in which the collection is defined or put together is irrelevant to its identity. Two definitions which happen to pick out the same elements are definitions of the same set. It is the presence of this axiom which makes it possible to prove that there can only be one empty set; any set which has no members has the same members, i.e. none, whether it is the set of numbers both greater than and less than 2 or the set whose only member is the greatest finite number.

The *null set axiom* assures that there is a set containing no elements. Moreover this is the only set whose existence is directly asserted. Every other set is constructed in some way or other from this set. This means that the universe of sets given by the ZF axioms is peculiarly a mathematician's universe. There are no individual objects, such as people, stones or trees, or collections of them in this universe. It is a wholly abstract universe generated, as it were, out of nothing.

The *pair set axiom* allows the first crucial step to be taken – to get, or assure the existence of, a set which has a member when the only set whose existence is otherwise assured is one with no members. For the null set is itself a set, an object, and we can thus form the set $\{\emptyset\}$, whose only member is the null set. So now there are two objects, $\varnothing$ and $\{\varnothing\}$, and the pair set axiom assures the existence of the sets $\{\{\varnothing\}\}$ and $\{\varnothing, \{\varnothing\}\}$, the latter containing two objects, so giving four objects in all. Applying the axiom again, we are assured of the existence of all sets of pairs of these four objects plus all sets containing each of them as their only member. Repeated applications of the pair set axiom thus assure the existence of any desired finite number of sets but each with only one or two elements.

The *sum set axiom* assures the existence of sets containing any desired finite number of elements by asserting the existing of the union of sets already defined. Thus,

$$\mathsf{U}\{\{\varnothing, \{\varnothing\}\}, \{\{\varnothing, \{\varnothing\}\}\}\} = \{\varnothing, \{\varnothing\}, \{\varnothing, \{\varnothing\}\}\}$$

i.e. the union of a pair of sets with no element in common, one containing two elements, the other containing one element, gives a set containing three elements. Larger sets result from repeatedly forming unions. But the sets will all be finite since only finitely many sets can be shown to exist by finitely many applications of the pair set and sum set axioms.

The *axiom of infinity* assures the existence of at least one infinite set, from which others can be generated, or proved to exist. It asserts the existence of a set to which $\varnothing$ belongs and which is such that if $\varnothing$ belongs to it then so also does $\varnothing \cup \{\varnothing\}$, the set whose members either belong to $\varnothing$ or to $\{\varnothing\}$. Since $\varnothing$ has no members, $\varnothing \cup \{\varnothing\}$ has just one member, namely $\varnothing$, and is thus identical with $\{\varnothing\}$. But then $\{\varnothing\} \cup \{\{\varnothing\}\} = \{\varnothing, \{\varnothing\}\}$ also belongs to the set, and so on. So the set asserted to exist contains a sequence whose initial segment is:

$$\{\varnothing, \ \{\varnothing\}, \ \{\varnothing, \ \{\varnothing\}\}, \{\varnothing, \{\varnothing\}, \{\varnothing, \{\varnothing\}\}\}, \ldots\}$$
$$\{0 \quad 1 \qquad 2 \qquad\qquad 3 \qquad\qquad \ldots\}$$

Here each element is a set which contains exactly one more element than that which preceded it. The set consisting only of this sequence (the smallest set which would satisfy the requirement of the axiom) is thus what gives a set-theoretic representation of the set of natural numbers, each number n being identified with the set in this sequence containing n members. Since the set itself is uniquely defined and forms a simply infinite sequence, well ordered by the membership relation, it is identified with the first infinite ordinal number ω. Given the way in which the sequence has been constructed, this identification means that each natural number n is identical with the set of numbers less than n and ω is identical with the set of ordinal numbers less than it (i.e. with the set of finite ordinal numbers, treated as being identical with the natural numbers).

These first five axioms are uncontroversial, and obvious in the sense that any set theory within which one wanted to be able to do even elementary arithmetic would have to contain them. The *axiom of foundation* is much less obvious and was one of the last axioms to be adopted. What it in effect says is that in any set there are elements which are minimal with respect to the membership relation; i.e. even when we have an infinite set S, S cannot contain an infinite sequence $X = \{x_i: i \in N \ \& \ \neg \ (x_i = \varnothing)\}$ of members such that $\ldots \in x_2 \in x_1 \in x_0$. For X would be a set which has an element in common with every one of its elements, since for each x_i, x_{i-1} belongs both to x_i and to X. Even sets like ω which are such that $0 \in 1 \in 2 \in 3 \in \ldots$ only contain, as members, sets which result from a finite number of applications of the operation of forming $x \cup \{x\}$ on $\varnothing$. This infinite set consists of *all* sets which can be formed by a finite number of applications of this operation but is not itself formed by an infinite number of applications of the operation. (Here we see an application of the strictures on a universe discussed in chapter 1; any universe of absolutely *all* things of a given kind will itself have to be a thing of a different kind, if it is to be regarded as a 'thing' at all.) The axiom of foundation rules out the formation of sets requiring a completed actual infinity of iterations of an operation of forming sets of sets. ZF uses only finite iterations together with the collecting together of all products of finite iteration. These collections can then form new starting points.

Similarly the axiom rules out the possibility of any set belonging to itself. For if $x \in x$ then $\{x\}$ would be a set which has an element in common with its only member and would thus be a set violating the axiom of foundation. The axiom thus plays a role in preventing the occurrence of paradoxes within the theory, for the set which is problematic in Russell's paradox, the set of sets which do not belong to themselves can be proved not to exist. No ZF sets belong to themselves, so the set of all sets not belonging to themselves, if there were one, would have to be the whole universe of ZF sets. But if this universe were itself a set it would belong to itself and so violate the axiom of foundation. So the universe of ZF sets is not a set and Russell's paradoxical set does not exist within the universe of ZF sets.

It is the axiom of subsets (sometimes called the axiom of separation) which makes it possible to collect together all those sets, contained within a set whose existence is already assured, which satisfy a condition expressible in the language of set theory. Since this language only contains one undefined expression (the membership relation), sets can only be described in terms of their structure, as given by the membership relation, their actual members, the relations of their members to other sets, or their own membership relation to other sets. This may seem like a very limited vocabulary. What is remarkable is the expressive power of this language of sets, and this is something which could only have been shown through the effort to achieve a formal axiomatization. The important feature of the axiom of subsets is the way in which it restricts this kind of formation of sets.

Cantor, and others, had assumed that one could always collect together into a set all the things satisfying a given, meaningful description (an assumption otherwise known as the unrestricted comprehension axiom). The axiom of subsets does not allow this, but only allows for the use of descriptions to create subsets of a set whose existence is already assured on the basis of the other axioms (and is thus a restricted comprehension axiom). In particular this means that, once the properties of being a cardinal number, and of being an ordinal number, have been defined it will not be possible to form the set of all ordinal or of all cardinal numbers, but only of those which are contained within some given set. The paradoxes of Cantor and of Burali Forti will thus be avoided. Similarly the axiom of subsets does not allow for the formation of the universe of sets as itself a set ($\{x : x = x\}$). The fact that such totalities, were they to be sets, would lead to contradictions is instead taken as a proof that they are not sets.

But the axiom of subsets did not prove sufficient to the mathematician's needs, for the mathematician is frequently not concerned with the characterization and classification of entities, or with the totalities which result. Since the seventeenth century much of the concern of mathematics has been with functions, transformations or mappings of various kinds. A function $f(x)$ is presumed to have been defined over a given domain D when it can be shown that for every x in D, $f(x)$ takes a value y and this value is unique, i.e.

$$\forall x(x \in D \Rightarrow \exists y[f(x) = y \And \forall z(f(x) = z \Rightarrow y = z)])$$

When a function is defined over a given domain D it is presumed that the values taken by f also form a set (the range of f), but this set will frequently not be a subset of the original domain D. The contrast between the two ways of forming sets from a given set can be pictured as in figure 6.1.

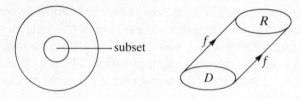

Figure 6.1

The *axiom of replacement* is a stronger axiom than the axiom of subsets in that it allows for both kinds of set formation whilst applying the same sort of restriction present in the subset axiom, i.e. it only allows for the formation of a set as the set of all values taken by a function f defined over a set whose existence is already assured by other axioms. Without this axiom it would be impossible to prove the existence of further ordinal numbers beyond $\omega + n$, for finite n. The other axioms assure the existence of $\omega \cup \{\omega\} = \omega + 1$, $(\omega \cup \{\omega\}) \cup \{\omega \cup \{\omega\}\} = \omega + 2 \ldots$ but not of any set to which they all belong. But since it is possible to define a function over ω which takes these as values, namely $f(n) = \omega + n$, the axiom of replacement assures the existence of a set to which all these values belong. The union of this set with ω then gives a representation of $\omega + \omega$. In a similar way all the ordinal numbers of the second number class will now have been provided for.

Strictly speaking neither the axiom of subsets nor the axiom of replacement is a single axiom. They are axiom schemata, i.e.

expressions which yield axioms when a specific description or relational expression, written in the language of ZF, is substituted for the place-holding variable expression. They could only be written out as, or.viewed as, axioms if they were expressed as universal generalizations over the properties of sets, or over the relations between sets (i.e. only if second-order quantification were used). These quantifiers would not be restricted to properties or relations expressible in the language of ZF, so their use would introduce a considerable indeterminacy into this axiomatic specification of the universe of sets. It would not be possible to investigate systematically the totality of sets required to exist by the axioms because this would not be fully controlled by them. The restricted and schematic (first-order) form of these axioms limits the sets required to exist by them (or formed by their means) to those definable in terms of sets already required to exist by the other axioms.

It is, however, possible to extend the language of ZF by introducing variables, X, Y, . . . , to range over classes, as opposed to sets in such a way that the only classes which exist are those classes of *all* sets (without restriction to being contained in some given set) which satisfy a given first-order formula expressible in the language of ZF. We could add the comprehension axiom

[CA] $\exists X \forall y (y \in X \Leftrightarrow Fy)$

where Fy is a formula expressible in the (first-order) language of ZF and which thus only contains quantification over sets, not over classes). Systems of set theory extended by an axiom of this kind have been extensively used by von Neumann, Bernays, Gödel and others and are described as 'predicative' extensions of set theory (a term to be further discussed in chapter 7). In such systems the axiom schemata of subsets and replacement can be rewritten as a single axiom

$z \exists y \forall x (x \in y \Leftrightarrow x \in \forall z \& x \in X)$, where X is a free class variable

It can, however, be shown that this form of extension, which makes for ease of expression, does not increase the power of the theory, i.e. there are no results about sets or set existence which can be proved in the extended theory which could not be proved in ZF. The thing which distinguishes classes from sets in such extended versions of set theory is that the classes are all classes of *sets*, and classes themselves are not allowed to be formed into further classes. Thus a class can never belong either to another class or to a set.

If, on the other hand, the 'F' in [CA] is allowed to contain bound class variables (as in Bernays–Morse set theory) the result is a system (an impredicative extension of ZF) which is stronger than ZF. It is, however, very difficult to justify the introduction of quantification over classes at the same time as drawing a sharp distinction between classes and sets which denies that classes can be formed into further classes. If quantification is throughout being treated extensionally, then to admit quantification over classes is to presume that the classes of sets form a determinate totality which ought itself to be admitted as a class, a class to which all classes of sets belong. It can, however, be shown that the strengthening of ZF achieved by admitting quantification over classes can also be achieved by the addition of an axiom written in the first-order language of ZF, and thus does not require the introduction of class variables. This axiom would assert the existence of a strongly inaccessible cardinal number (one which, roughly speaking, cannot be reached from below either by adding a smaller number of smaller cardinals together or by taking 2^α or $\aleph_\alpha$, for smaller cardinal numbers α). Issues relating to these points will be taken up in chapter 8.

The concerns voiced above about the underspecification of the universe of sets introduced by full second-order axioms might seem to be misplaced in the light of the power set axiom, for this asserts the existence, for any set a, of the power set $P(a)$ of a, which is the set of *all* subsets of a. Here there are no restrictions given as to the means by which subsets are defined or constructed. There is nothing which says that the only members of $P(a)$ are those which can be defined by means of expressions written in the language of ZF. So although the existence of $P(a)$ is assured by the axioms, its exact membership is underdetermined by them. The axiom is, however, important for, without it, it would be impossible to prove the existence of any non-denumerable sets and hence of any ordinal numbers which do not belong to the second number class. It was by showing that, for any set a (whether finite or infinite), the cardinality of $P(a)$, its power set, must be greater than the cardinality of a that Cantor was led to think it possible to extend the notion of number into the infinite, for it suggests that not all infinite sets are identical in respects which are something like number.

With the inclusion of the power set axiom it becomes possible to prove the existence of the second number class as a set, whereas, without it, it is only possible to prove the existence of all the members of that class. It is then also possible to prove that the cardinality of this set is greater than that of ω and that there can be no set with a cardinality greater than that of ω but less than that of the second number class. So the second number class is then identified with, or used as the set theoretic representation of the first non-denumerable ordinal number, ω_1. In addition, the cardinal number of a set s is defined as being the least ordinal number α such that there is a one–one correspondence between s and α. This means that since ω is the smallest infinite ordinal number, $\aleph_0$ will be identified with ω, and since ω_1 is the smallest non-denumerable ordinal number, $\aleph_1$ will be identified with ω_1. Thus infinite cardinal numbers are identified with ordinal numbers of a special kind (initial ordinals), those which give the ordinal number of a number class and which have now been identified with the number class itself given in its natural ordering.

The axiom of choice is the most controversial of the axioms listed here. It is not treated as a basic axiom and its use in any proof is always explicitly noted, so that a distinction is drawn between results which can be proved without appeal to it and those which have only been proved with its help. In the context of ZF, it can be proved to be equivalent to the assertion that given any set a there is a relation R which is a well-ordering of a. In papers published in 1904 and 1908, Zermelo proved this assertion, which was thus known as the well-ordering theorem. It was by examining the principles appealed to in his proof that Zermelo was led to formulate the axiom of choice, which he saw as a basic logical principle. It was indeed a principle to which mathematician's had implicitly appealed in other contexts (for example, in proving the equivalence of the two standard ways of defining a continuous function in analysis). Another context in which appeal needs to be made to this axiom is in proving the equivalence of two definitions of 'finite' (or 'infinite').

A set S is *ordinary finite* if it is empty or there exists a natural number n such that there is a one–one correspondence between S and n. A set S is *ordinary infinite* if it is not ordinary finite.

A set S is *Dedekind infinite* if there is a proper subset $S°$ of S such that there is a one–one correspondence between S and $S°$. A set S is *Dedekind finite* if it is not Dedekind infinite.

Suppose S is an ordinary finite set. Then there is a natural number n such that there is a one–one correspondence f between n and S. Suppose S were not Dedekind finite, then there would be a one–one correspondence g between S and some proper subset $S°$ of S. The restriction of f to $S°$ will be a one–one correspondence between $S°$ and some proper subset of n. But then there would be a one–one correspondence between n and a proper subset of itself (for any i in n, $f(i)$ is in S, $g(f(i))$ is in $S°$ and $f^{-1}(g(f(i))$ is in $n°$, and f, g, f^{-1} are all one–one correspondences). But it can be shown by induction that there can be no one–one correspondence between a natural number and a proper subset of itself.

Suppose S is not an ordinary finite set, i.e. is ordinary infinite. Then it can be shown that S contains a subset $S°$ which can be put in one–one correspondence with the set of natural numbers, and since the set of natural numbers is Dedekind infinite, the set S will be Dedekind infinite.

If S is not ordinary finite, then it is not empty, so we can select x_0 from S. Then $S - x_0$ will not be empty for otherwise $S = \{x_0\}$ which is finite. So we can select x_1 from $S - \{x_0\}$, and $S - \{x_0, x_1\}$ will not be empty. Similarly for any n, $S - \{x_0 \ldots x_n\}$ will not be empty and it will be possible to select x_{n+1} from it. The set T consisting of all these selections must therefore be contained in S and can be put in one–one correspondence with the whole sequence of natural numbers. The correspondence defined by mapping x_n to x_{n+1} *for* x_n in T and mapping every other element of S (if there are any) to itself will be a one–one correspondence between the whole of S and $S - \{x_0\}$.

It is in assuming that such an infinite sequence of selections can be made and the results collected into a set that an appeal is made to the axiom of choice. The problem here is that whilst it would clearly be quite legitimate to base an argument on the assumption that a finite sequence of choices can be made, it is not clear that it is legitimate to extrapolate this to an infinite sequence. No one could actually make infinitely many choices. It might be claimed that it is unnecessary anthropomorphizing to imagine a set arising from a sequence of choices; the question at issue is, rather, one of set existence. This is a point at which the more constructivist view of the role of the axioms and the view of the axioms as saying what exists independently of our constructions part company. The worries about the axiom of choice arise from the more constructivist position, for the set asserted to exist is one which we may have no means of constructing or specifying in any way. The axiom merely asserts that there is a set (a selection set) which contains just one element from each member of a given set S of non-empty, disjoint sets without giving any indication of how the selection is, or might be made. But from the point of view of the working mathematician there is no reason to discard the axiom, and there is reason to include it, given that it embodies a principle of reasoning which was antecedently in use. There is no reason to discard the axiom because in 1940 Gödel proved that the axiom of choice is consistent with the remaining axioms of ZF so its use cannot lead to any contradictions which would not have arisen without it. On the other hand, proofs which make use of the axiom are not maximally informative. Wherever possible it is desirable to have a constructive proof of the existence of a set because this means that one thereby has some information about the nature of the set – information based on the way it was 'constructed' and so shown to exist. It is for this reason that proofs which avoid appeal to the axiom of choice are sought wherever possible. It is also possible to, as it were, grade the assumptions made in terms of the 'number' of choices which need to be made. In the proof to show the equivalence of the two notions of 'finite', only a denumerable sequence of choices was required. So a weak form of the axiom of choice would have sufficed (an axiom of denumerable choice) – one which holds for denumerable sets of non-empty disjoint sets. Without even this form of the axiom of choice one would have to admit the possibility that there might be a set S which is ordinary infinite but not Dedekind infinite. This S would be such that there is no natural number n such that there is a one–one correspondence between S and n, and S would not contain any subset $S°$ which could be put in one–one correspondence either with the whole set S or with the whole set N, of natural numbers. S would then be incomparable, in respect of cardinality, with the set of natural numbers.

With the inclusion of the axiom of choice it becomes possible to prove both that every set can be well-ordered and to prove that any two sets A and B are comparable in respect of cardinality.

3 Transfinite Numbers in ZF

In explaining the role of the ZF axioms we have at the same time shown, in an informal way, how these provide for the existence of a well-ordered sequence of sets which can, by stipulative definition, be identified with Cantor's (transfinite) ordinal number sequence. The sequence of sets

$$\varnothing \quad \{\varnothing\} \quad \{\varnothing, \{\varnothing\}\} \quad \{\varnothing, \{\varnothing\}, \{\varnothing, \{\varnothing\}\}\} \cdots$$
$$0 \quad\quad 1 \quad\quad\quad 2 \quad\quad\quad\quad 3 \quad\quad\quad\quad\quad \cdots$$

is identified with the sequence of natural numbers. Here it should be noted that the sequence of sets used is well-ordered by the membership relation '$\in$', since $n < m$ if, and only if, $n \in m$. This idea can be generalized and formalized in such a way as to give a characterization of ordinal numbers which is expressible in the language of ZF and which is independent of any talk of principles of generation:

Definition A set a is *transitive* if, and only if,

$$\forall x \forall y (x \in a \ \& \ y \in x \Rightarrow y \in a)$$

i.e. a is transitive if, and only if, whenever x belongs to a, all the members of x also belong to a (hence $x \in a$ if, and only if, $x \subset a$).

Definition A set α is an *ordinal number* $(On(\alpha))$ if, and only if, α is well-ordered by '$\in$' and is transitive.

This could have been written as a fully formal definition in the language of ZF since both the conditions for being well-ordered by '$\in$' and being transitive can be expressed as first-order formulae in which '$\in$' is the only non-logical primitive.

It can then be proved, from the ZF axioms without the axiom of choice that:

1 For any well-ordered set a, there is a unique α such that $On(\alpha)$ and α is order-isomorphic to a.
2 If two ordinal numbers are order isomorphic, they are identical. This means that amongst the ordinals there will be exactly one representative of the order type of any given well-ordered set and that it is therefore legitimate to talk of *the* ordinal number sequence, and of *the* ordinal number of a well-ordered set.
3 For any ordinal number α, $\alpha = \{x : On(x) \ \& \ x \in \alpha\}$, i.e. any ordinal number is identical with the set of all ordinal numbers less than itself.

Definitions An ordinal number α is a *successor ordinal* if, and only if,

$$\exists x (On(x) \ \& \ \alpha = x \cup \{x\})$$

Since any number has a successor and can have at most one successor we can define $\alpha + 1 = \alpha \cup \{\alpha\}$.

An ordinal number α is a *limit ordinal* if, and only if, $\neg \, (\alpha = 0)$ and α is not a successor ordinal.

An ordinal number α is a *natural number* if, and only if, every ordinal number β such that $\beta \leqslant \alpha$ is either a successor number or $\varnothing$, i.e. $\forall x (x \subseteq x \Rightarrow \exists y (x = y \cup \{y\} \lor x = \varnothing))$.

The axiom of infinity can then be interpreted as saying that a limit ordinal exists, and that ω can be introduced to stand for the set of natural numbers. ω will then be the smallest limit ordinal. Cardinal numbers can then be identified with special ordinal numbers – those which are initial ordinals.

Definition An ordinal number α is an *initial ordinal* if, and only if, for every ordinal number $\beta < \alpha$ the cardinality of β is less than that of α.

Because the ordinal numbers form a well-ordered sequence, it is assured that amongst those having the same cardinality there will be a least. So by this means it is possible to select just one ordinal number to represent all those having the same cardinality. For example:

1 Every natural number will also be a cardinal number, since for each natural number n, there is just one ordinal number having that finite cardinality.
2 ω will be the first infinite cardinal number (denoted by $\aleph_0$) and $\aleph_{\alpha+1}$ will be used to denote the least ordinal of cardinality greater than $\aleph_\alpha$.

It can be proved that, for each ordinal number α, there does exist an ordinal of greater cardinality, i.e. that the cardinal number sequence does extend indefinitely. This is proved by considering the set of all well-orderings on α (a subset of the set of all sets of ordered pairs of members of α). Each well-ordering of α will have an ordinal number of the same cardinality as α. The union of the set of all ordinal numbers of well-orderings of α can be shown to be an ordinal number whose cardinality is greater than α and to be the least such ordinal number. $\aleph_\alpha$ can then be defined as the least infinite cardinal number which is greater than $\aleph_\beta$, for every $\beta < \alpha$. This means that $\aleph_{\alpha+1}$ is the least infinite cardinal greater than $\aleph_\alpha$, $\aleph_\omega$ is the least infinite cardinal greater than all of $\aleph_0, \aleph_1, \aleph_2, \ldots$, and so on.

Given these definitions Cantor's continuum hypothesis becomes a precisely statable proposition of ZF. It says:

[CH] $C(P(\omega)) = \omega_1$, or $2^{\aleph_0} = \aleph_1$, since $C(P(\omega)) = 2^{\aleph_0}$ and $\omega_1 = \aleph_1$.

The generalization of this, the generalized continuum hypothesis is then:

[GCH] $C(P(\aleph_\alpha)) = \aleph_{\alpha+1}$ or $2^{\aleph_\alpha} = \aleph_{\alpha+1}$.

The continuum hypothesis has now been given a precise sense. It is a statement in the language of ZF about sets which can be proved to exist from the axioms of ZF. The available means of proof or disproof for the continuum hypothesis have also been made explicit with the axiomatization of set theory. However, once the continuum hypothesis had been made precise in this way it also became possible to prove that it could neither be proved nor disproved by reference to the ZF axioms. First Gödel (1938) showed that both it and the axiom of choice are consistent with the ZF axioms, and then Cohen (1964) showed that its negation and the negation of the axiom of choice are also consistent with the ZF axioms. These proofs are conducted by providing different models of the ZF axioms which will be discussed in chapter 8. It turns out that the ZF axioms put very little restriction on where $2^{\aleph_0}$ should be located on the ordinal number scale. The restriction is merely that $2^{\aleph_0}$ cannot be equal to $\aleph_\mu$ if μ is the limit of a strictly increasing ω-sequence of ordinals.

Thus we have an apparently precise question which also apparently has no precise answer. This is the situation referred to in the Introduction and it is one which cannot be interpreted without raising the sort of philosophical questions about the status of the axiomatization of set theory expressed there. Before tackling these questions, however, it is necessary to look at the alternative and more directly philosophical response to the contradictions which arose in Cantor's set theoretic treatment of numbers.

7

Logical Objects and Logical Types

Cantor was led to venture into the transfinite by his attempts to further the project of arithmetizing analysis. This project was itself motivated by a concern to put analysis on a secure footing and to remove the need for geometrical intuition. The definition of pathological functions (ones which defy pictorial representation) had shown geometrical intuition to be an inadequate basis for resolving questions concerning the behaviour of functions, the correct characterization of continuity, etc. Cantor's naive set theory, however, replaced geometrical intuition by an appeal to intuitions about sets, without giving any reason for thinking this to be ultimately more reliable or mathematically adequate than geometrical intuition. The set-theoretic paradoxes rendered claims about the superior reliability of set-theoretic intuition doubtful in the extreme.

1 Frege, Logic and Arithmetic

It was Frege's view that in arithmetic there need be no appeal to any sort of intuition, for arithmetic reduces ultimately to logic in the strong sense that (a) arithmetical objects (numbers) are logical objects, and (b) all true statements about these objects can be proved from definitions by appeal only to logical laws. This claim covers both statements of pure arithmetic, such as '2 + 2 = 4', and statements in which numbers are applied, such as 'If in the basket there are two apples and four pears, then there are two more pears in the basket than there are apples'. He did not, however, think that intuition could be dispensed with throughout mathematics. He held that geometry, because it concerns the forms of space and time, cannot be divorced from intuition.

Since Frege accepted Cantor's basic idea that the concept of number should be elucidated by reference to sets, his claims about arithmetic amount to the claims that (c) the theory of sets can be reduced to the theory of classes, and (d) classes are logical objects in that the theory of classes can be developed as a pure logical theory. His concern in making such a claim was primarily philosophical. It was a concern to reveal the foundations of arithmetic by providing a precise analysis of the concept of number, an analysis which would at the same time reveal the nature of arithmetical truths by showing how they can be established. If his philosophic account of arithmetic were correct, it would provide a guarantee of the consistency of the arithmetic of finite numbers and would also legitimize its extension into the transfinite, for Frege did not restrict 'number' to the finite.

That the theory of classes should be nothing other than logic is suggested by the long-standing association between logic and classes as the extensions of terms, which was discussed in chapter 3. But as was noted in that chapter, the notion of class associated with that tradition is far from being precise. It was ambiguous between definite collections of actual objects and indefinite extensions of terms taken over all possible objects. In particular, the extensionalist approach runs into problems with the null set and the universe, as was seen in that chapter. It is just this ambiguity which suggests that there is no distinction to be drawn between a theory of classes and a theory of sets.

It had, however, been clear since the seventeenth century, with the development of algebra and the emergence of 'function' as a central mathematical notion, that traditional Aristotelian logic was quite incapable of representing the kind of reasoning being employed in mathematics. The logic of terms and their relations provides no effective framework for formalizing reasoning involving relations. Frege's achievement was to extend formal logic in such a way that it became capable of handling relations and functions. In fact Frege brought about this extension of logic by drawing on the mathematical notion of a function. His first move was to replace the notion of a term, which is something that can occur in either subject or predicate position in a sentence, by the pair of contrasted notions 'concept' and 'object'. Here 'Socrates is a man', instead of being treated as a relation between terms (SaM) is treated as the application of the concept 'is a man' (Mx) to the object Socrates (s), yielding a value (a truth value) denoted by the sentence 'Ms'. A concept, then, is a function of a single variable whose values are always truth values (the True or the False).

The difference between concepts and objects, like the difference between numbers and functions, is a difference in kind. Concepts, like functions, are incomplete, having within them a place for completion by an object (this being marked by the place-holding free variable 'x'). Objects, on the other hand, are complete, self-subsistent entities. Once this analogy is accepted, it is clear that relations can be treated in a parallel way by analogy with functions of more than one variable. A relation such as 'is less than' (Lx, y) takes pairs (or, more generally, n-tuples) of objects as arguments and has a value which is always one of the two truth values. Thus $L3, 5 = $ True and $L5, 3 = $ False. Furthermore, it then becomes possible to treat mathematical functions as relations; $fx = y$ can be seen as a two-place relation, a function whose value is a truth value. So it seems that it would be possible in fact to reduce all functions to functions whose values are truth values, and if such functions can be fully handled within formal logic, then the reduction of arithmetic to logic looks at least plausible.

However, this innovation on its own does not significantly extend the power of logic. It is important because it paved the way for Frege's other major innovation – the introduction of the now familiar quantifier and bound variable notation. Here again Frege drew on a mathematical analogy. Operators such as the definite integral

$$\int_a^b fx \, dx$$

convert an expression for a function into one which denotes a numerical value, where the numerical value is dependent on all of the infinitely many values taken by fx for arguments in the specified range. When 'x' occurs in such an expression it is no longer a free variable, indicating an incompleteness to be filled in by application to an object. It is a bound variable, a device for referring to aspects of the graph of a function which depend on the values taken by the function over a whole range of arguments. Transferring this to concepts, it suggests that a similar device could be used to express those statements which were previously interpreted as expressing relations between the extensions of terms. The extension of a term M can now be seen as the class of objects for which the concept Mx takes the value True ($\{x: Mx\}$).

Since concepts can only take one of the two truth values, the only sort of statement which can be made about their values over a given range of arguments will be those depending on the distribution of Trues and Falses in this range, and this is what quantifiers make possible. $\forall x Mx$ takes the value True or False depending on whether there is or is not an object for which Mx takes the value False. The truth value of $\exists x Mx$ depends on whether there is an object for which Mx takes the value True. These are clearly the simplest possible cases. Numerically definite quantifiers give more information, but it can be shown that they can be defined. For example,

$$\exists_1 x Mx \equiv_{df} \exists x (Mx \ \& \ \forall y (My \Rightarrow x = y)).$$

This gives the new notation the power to express all that could be expressed in the traditional symbolism for syllogisms, giving the now usual formulations:

All S are P	$\forall x \neg (Sx \ \& \ \neg Px)$ or $\forall x (Sx \Rightarrow Px)$
No S are P	$\forall x \neg (Sx \ \& \ Px)$
Some S are P	$\exists x (Sx \ \& \ Px)$
Some S are not P	$\exists x (Sx \ \& \ \neg Px)$

Here, rather than comparing the extensions of two terms, a complex concept is formed, e.g. $\neg (Sx \ \& \ Px)$, and then a statement made about whether this takes the value True for any object as argument. This closely parallels the procedure used in Venn diagrams and in Leibniz's algebraic notation (see pp. 37–8), whilst at the same time it effects a considerable clarification of the issues surrounding the notion of a class (even though it could not be said to result in a wholly clear notion).

So far the impact of Frege's introduction of the quantifier and bound variable notation has been considered only in respect of the way in which it provides for a reformulation of syllogistic logic and the notions traditionally associated with it. However, the quantifiers only become really significant when they are employed in connection with relations. For, again drawing on the analogy with mathematical functions, operators which bind one variable in a function of several arguments act as higher level functions mapping functions of n arguments to functions of $n - 1$ arguments. For example, when a quantifier is applied to a two-place relation, the result is a concept. From a relation Rxy the following concepts can thus be defined: $\exists xRxy$, $\forall xRxy$, $\exists yRxy$, $\forall yRxy$. If Rxy says that x and y are natural numbers and $x < y$, then $\exists xRxy$ says that y is a natural number such that there is a natural number less than y, which would be true for all numbers other than 0; $\exists yRxy$ says that x is a natural number such that there is a natural number greater than x, which would be true for all natural numbers (since there is no greatest); $\forall xRxy$ says that y is a natural number greater than all natural numbers, which would be false for all natural numbers; $\forall yRxy$ says that y is a natural number which is less than every natural number, which would be false for every natural number because no natural number is less than itself.

Thus the introduction of the quantifier and bound variable notation, together with the treatment of relations by analogy with functions of two or more arguments, greatly extends the means for defining concepts. (The extension effected is closely related to that extension in the power of ZF which is effected by the addition of the axiom of replacement.) Specifically it now becomes possible to define concepts from relations, where these concepts depend on the 'structural' characteristics of the 'graph' (value range, or extension) of the relation involved. In addition it becomes possible to characterize the 'structural' properties, such as transitivity of relations themselves (a relation is transitive if, and only if, $\forall x\forall y\forall z(Rxy \ \& \ Ryz \Rightarrow Rxz)$). And, as we have seen, many of the problems preoccupying Cantor and others working on the foundations of mathematics revolved around attempted characterizations of structures. Thus it was not by any means an unreasonable hypothesis to suggest that mathematics, in so far as it is concerned to study the possible structures on domains of individual objects, might be reduced to logic, in Frege's extended sense of logic. As Frege himself points out (Frege, 1959, p. 100), he has so extended the power of defining concepts that for him 'analytic truth' (following from definitions by application of the laws of logic alone) no longer means what it did for Kant, who was working within a basically Aristotelian framework (assuming that conjuction and negation of terms (concepts) are the basic means of defining complex terms (concepts)). In the pictorial terms of Venn diagrams, Kant would be restricted to defining concepts from concepts and the concepts definable from two given concepts correspond to the sixteen regions having boundaries in the Venn diagram of figure 7.1.

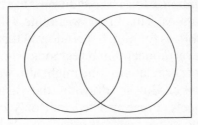

Figure 7.1

Frege, however, can define concepts without there being any primitive concepts at all by starting with a primitive relation. This goes beyond what is easily represented spatially. An analogy would be with the way in which a number of different two-dimensional shapes can be produced as the shadows cast by a single three-dimensional object onto planes oriented at different angles to it. A demonstration of the power thus introduced is provided by the language of ZF, which has the single non-logical primitive '$\in$'.

However, this is not enough to establish Frege's claims about arithmetic, for his claim that arithmetic reduces to logic requires it to be shown that all arithmetic concepts and objects (including the concept of 'number' and the individual numbers) can be defined in purely logical terms. Given Cantor's definition of cardinal numbers, the first step on this road would be to express the condition 'there is a one–one correspondence between the Fs and the Gs' in purely logical terms. Then it would be possible to give a purely logical expression of 'the (cardinal) number of Fs is the same as the (cardinal) number of Gs'. This can be done in stages:

R is a *one–one correspondence* between Fs and Gs, $\mathscr{C}(R, F, G)$ iff

1 R is a *one–one correspondence*, i.e.

$$\forall x \forall y \forall z ((Rxy \text{ \& } Rxz \Rightarrow y = z) \text{ \& } (Rxy \text{ \& } Rzy \Rightarrow x = z))$$

2 Every F is R-related to some G and vice versa, i.e.

$$\neg \exists x (Fx \text{ \& } \neg \exists y (Rxy \text{ \& } Gy)) \text{ \& } \neg \exists y (Gy \text{ \& } \neg \exists x (Rxy \text{ \& } Fx))$$

the number of Fs = the number of Gs iff $\exists R \mathscr{C}(R, F, G)$

However, to symbolize this last condition it has been necessary to quantify over relations. This Frege was quite willing to do, but it posed certain problems for him. If it is allowed that there are higher order (or higher level) concepts and relations, whose arguments are first-order concepts or relations, one might also want to know what their extensions are. But in drawing the distinction between concepts and objects, Frege insisted that only objects could fall under concepts and hence that only objects could belong to classes, which would in turn be objects. How, then can either concepts or relations now be allowed to fall under further concepts or relations, or be allowed to form classes? Frege's answer relies on the assumption that logic is extensional, and that concepts are to be regarded as identical when they have the same extensions. This means that, given any statement, $\mathscr{G}(F)$, about a concept F, its truth value will remain unaltered when another concept expression, say G, is substituted for F provided that F and G are extensionally equivalent (i.e. that $\forall x(Fx \equiv Gx)$). In this case it could be said that any higher order concept or relation must be one expressing a characteristic of, or a relation between, concept extensions. We have seen the sense in which this would be true for those higher level concepts and relations which can be expressed by purely logical means. The quantifiers can be viewed as second-order concepts—concepts which yield a truth value for a concept as argument – but they can clearly also be interpreted as saying something about the extensions of concepts. If the expression $\{x: Fx\}$ is introduced to denote the extension of the concept Fx and the membership relation '$\in$' is defined as a relation between objects such that

$a \in \{x: Fx\}$ iff   *Fa* and
$a \in b$ is false whenever b is not a class

(i.e. there is no concept of which it is the extension), then $\exists x(x \in y)$ will be a concept which has all non-empty classes in its extension. Moreover,

$$\exists x(x \in \{y : Fy\}) \quad \text{iff} \quad \exists xFx$$

So, on the assumption that all classes are the extensions of concepts, which has been built into the definition of the membership relation, $\exists x(x \in y)$ will be a first-order concept having an extension which contains the extension of every concept Fx for which $\exists xFx$ is true. $\exists x(x \in y)$ can thus do duty for the second-order concept $\exists x\phi x$. In this way Frege is able to obtain the effect of quantification over concepts and of having higher order concepts and relations but without having to make really essential use of them, since he can argue that they can all be reduced back to first-order concepts and relations via the membership relation.

But, as Russell showed, the combination of assumptions required to effect this reduction is enough to render Frege's theory of classes inconsistent. To show this one can start from Frege's definition of the membership relation:

$$
\begin{array}{llll}
x \in y & \text{iff} & \exists \phi(y = \{x : \phi x\} \,\&\, \phi x) & \text{so} & (1) \\
x \in x & \text{iff} & \exists \phi(x = \{x : \phi x\} \,\&\, \phi x) & \text{and} & (2) \\
x \notin x & \text{iff} & \neg\exists \phi(x = \{x : \phi x\} \,\&\, \phi x) & & (3)
\end{array}
$$

If every concept has an extension, then the concept denoted by the expression '$x \notin x$' must also have an extension and we are entitled to define a class

$$R = \{x : x \notin x\} = \{x : \neg\exists \phi(x = \{x : \phi x\} \,\&\, \phi x\}$$

R will then be the class of objects which are either not classes or are classes which do not belong to themselves. Since R itself is an object, it must be the case either that it belongs to itself or that it does not belong to itself.

Suppose $R \notin R$, then from (3) $\neg\exists \phi(R = \{x : \phi x\}$ and $\phi R)$ Thus R satisfies the defining condition for membership of R, i.e. $R \in R$, which contradicts our supposition.

So suppose $R \in R$, then from (2) $\exists\phi(R = \{x:\phi x\} \& \phi R)$
So suppose $R = \{x: Fx\} \& FR$, then we have

$$\{x: Fx\} = R = \{x: \neg\exists\phi(x = \{x:\phi x\} \& \phi x\})$$

from which it follows that

$$FR \quad \text{iff} \quad \neg\exists\phi(R = \{x:\phi x\} \& \phi R)$$

Since we have supposed FR to be true, $\neg\exists\phi(R = \{x:\phi x\} \& \phi R)$ must also be true. But this contradicts the assumption that $R = \{x: Fx\} \& FR$.

So both $R \in R$ and $R \notin R$ lead to contradictions.

The assumptions used in the derivation of this contradiction are as follows:

$$\{x: Fx\} = \{x: Gx\} \quad \text{entails} \quad \forall x(Fx \equiv Gx)$$
$$x \in y \quad \text{iff} \quad \exists\phi(y = \{x:\phi x\} \& \phi x)$$

These together mean that

$$a = \{x: Fx\} \quad \text{entails} \quad \forall x(x \in a \equiv Fx)$$

And since it is assumed that for every first-order concept Fx there is an object, $\{x: Fx\}$, which is the extension of the concept, this amounts to assuming what has been called the axiom of unrestricted comprehension, which says that every (first-order) concept has an extension, i.e.

$$\forall\phi\exists x(x = \{y: \phi y\} \& \forall y(y \in x \equiv \phi y))$$

It is clear, therefore, that these assumptions are not mutually consistent, and that unrestricted comprehension cannot be assumed. But what has gone wrong?

2 Frege's Universe

Before turning to Russell's answer to this question it is worth considering the character of Fregean classes in more detail. Frege argues strongly in favour of treating classes as the extensions of concepts, rather than as aggregates or arbitrary collections of objects. That is, he distinguishes the notion of 'class' from that of 'set' as given by the explications of Cantor and Hausdorff (pp. 98–9). He argues that it is the notion of 'class' rather than that of 'set' or 'aggregate' that is required to form a coherent foundation for arithmetic. This is because (a) the null class, necessary for the definition of 0, makes sense as the extension of a concept under which nothing falls (a contradictory concept), whereas an empty collection, or aggregation, of objects makes no sense; (b) objects are not indepedent of concepts in that what is to be counted as *one* object depends on the specification, given by a concept of the *kind* of objects to be counted. Faced with a page of print one cannot say how many objects there are on it. One needs to know whether to count letters, words, sentences, lines, etc. Numbers can, therefore, only be applied to the extensions of concepts, for it is concepts which tell us how to count the objects which fall under them by determining the unit of counting. Even an aggregate, to be an aggregate of *objects*, a determinate plurality, will have to be made up of objects of given kinds. It is true that a finite collection S can be specified simply by listing its members ($x \in F$ iff $x = a \lor x = b \lor x = c$), where these might be of very different kinds. In this case there would be no obvious single concept under which they all fall. But Frege would argue that each of the names used itself has a sense which includes a specification of the kind of object named. Thus a membership condition given as a finite list of members will itself constitute a disjunctive concept which determines how to count the things falling under it. This does, of course, leave open the question of how to regard infinite sets. Are there infinite disjunctive concepts to correspond to infinite lists? If there were, then almost anything which Cantor would recognize as a set would also be plausibly thought to be a Fregean class. But if the powers of definition of concepts do not extend to infinite disjunction and the existence of concepts is strictly co-ordinate with the powers available for defining them, then the Cantorian universe will contain sets which are not Fregean classes. So Frege's insistence that numbers can be applied only to the extensions of concepts makes the question of what concepts there are a crucial one for the foundations of mathematics.

This insistence also makes it possible for Frege to clarify the distinction between the membership relation and the part–whole, or inclusion, relation for classes. Traditional logic tended to confuse these because a singular statement (Socrates is a man) had to be treated as a universal statement in which 'Socrates' was a term applicable to just one thing. Thus, in the formal logic there was no difference between saying that Socrates belongs to the class of men and saying that he is part of, or is included in, the class of men. Nor was there any reflection of the difference between Socrates and the class of which he is the only member. With Frege's notation, however, there is a clear distinction between Ms and $\forall x \neg (Sx \ \& \ \neg Mx)$. In the first case 'Socrates' is treated as a name for an object and in the second 'is Socrates' is treated as a concept under which only one thing falls. So the sharp distinction between object and concept is reflected in the distinction between membership and inclusion relations; only an object can be a member of a class, and only a class can be included in another class. In addition, it leads to a distinction between an individual object and the singleton class, whose only member is that object. Again Frege argues that this distinction makes no sense if classes are thought to be aggregates of objects (see also Goodman (1964)). What Frege makes clear is that even if the criterion of identity for classes is

$$\{x: Fx\} = \{x: Gx\} \quad \text{iff} \quad \forall x (Fx \equiv Gx)$$

so that the identity of a class is determined by its members, rather than by any particular way in which a concept expression picks them out, this does not mean that a class is no more than its members. It is a new object, and there is a real distinction between a and $\{a\}$.

Frege's logic is extensional in that he treats concepts as identical when they have the same extensions (or as he calls them, 'value ranges'). Here again Frege draws on the analogy with mathematical functions. Treating concepts as identical when they have the same extensions is like treating functions as identical when they have the same graph. (Because Frege draws explicitly on this analogy, his value ranges are not quite identical with classes. The value range of a concept is more like a class of ordered pairs.) What this extensional treatment of concepts means is that whenever $\forall x (Fx \equiv Gx)$ is true it should be possible to substitute 'Gx' for 'Fx' in all contexts without altering the truth or falsity of what is said. 'Fx' and 'Gx' would thus be regarded as different expressions for the same concept – they have the same reference but different senses. Frege does not, however, treat concept expressions as standing for, or denoting, their extensions, for this would be to forget the distinction between concept and object. A class, which is an object, cannot be the reference of a concept expression; a concept expression stands for a concept. The extension of a concept is an object in its own right and is denoted by an expression such as $\{x: Fx\}$.

Because Frege insists on the priority of concepts over classes, his treatment of the standard cases of categorical statements ('all S are P', etc.) does not fit tidily into either the essential or the existential interpretations discussed in chapter 3. In his early works Frege certainly does not think of $\forall xFx$ as being just the conjunction of its instances; it is a statement which derives its sense and the determination of its truth value from the concept Fx. Frege also admits the independent existence of concepts and relations along with that of individual objects and thus departs from nominalism. On the other hand he, like the nominalists, had no time for distinctions between essential and accidental, potential and actual, but his universe of individual objects is not that of the nominalist. His actual universe is the universe of all objects and this includes many abstract objects (objects of thought) and, crucially, it includes classes. It was Frege's view that modal statements (it is necessary that ..., it is possible that ...) merely reflect subjective attitudes; they reflect the manner in which a person knows or believes a statement; there are no modalities in the world. So he cannot make a distinction between the actual universe and possible universes and consequently cannot distinguish between the extension of a concept Fx as the class of all actual Fs and the extension of Fx as the class of all possible Fs. This is important for the claim that classes are logical objects and that logic can yield arithmetic. For it is quite crucial to these claims that the denotation of an expression such as $\{x: Fx\}$ should be uniquely determined, and not be ambiguous between two different extensions.

It is far from clear, however, that Frege genuinely succeeds in removing this ambiguity. He has no room in which to recognize it, but nevertheless his view of the relation between concepts and counting pulls in two incompatible directions, and a stipulative unification cannot by itself render them compatible. Since concepts determine how to count the objects which fall under them, the grasp of a concept Fx must clearly include a grasp of what it is to be an object c for which Fc is True. Grasp of a concept can thus form the ground of judgements about all possible Fs in the manner required for non-extensional readings of universal statements (which was also required by the classical finitist). Here Frege seems to treat concepts in general as functioning like Aristotelian kind terms.

However, Frege also insists that every concept is defined over the whole universe of objects (i.e. takes either the value True or False for *any* object as argument) and the negation, $\neg Fx$, of any concept Fx, is also a concept which is such that $\neg Fa$ is True, if, and only if, Fa is False. So the extension of $\neg Fx$ is the complement, relative to the whole universe of (possible) objects, of Fx. Yet, for the reasons discussed in chapter 2, it is not reasonable to suppose that, by negating a concept which carries with it criteria of identity and individuation for the objects falling under it, we thereby get a negated concept which also carries with it criteria of identity and individuation for all the objects which fall under it, since these, being all the rest of the objects in the universe, must be very heterogeneous. By knowing how to count the number of words printed on a given page, I do not thereby know how to start counting all the things which are not words on that printed page. So if Frege's original argument for the need to treat classes as the extensions of concepts, rather than as aggregates, when analyzing notions of number is correct, it will tell against allowing negated concepts to be included as concepts to which notions of number apply. The logical notion 'concept' (explicated as a function from object to truth values), which will count negated concepts as concepts, does not therefore seem ideally suited for the analysis of number unless the argument about the interdependence between concepts and the objects falling under them is dropped. By insisting that it is to the extensions of concepts, including those formed by negation, that numbers, strictly speaking, have application, Frege would be committed to thinking that talk of the number of things which are not words on a given printed page makes sense. He does not explicitly make this general commitment, but says:

Only a concept which isolates what falls under it in a definite manner and which does not permit any arbitrary division of its parts, can be a unit relative to a finite Number.

and

Not all concepts possess this quality. We can, for example, divide up something falling under the concept 'red' into parts in a variety of ways, without the parts thereby ceasing to fall under the same concept 'red'. To a concept of this kind no finite number will belong. (Frege, 1959, p. 66)

This last remark leaves it open as to whether an infinite number will belong to the concept 'red', or whether this is a concept to which no number belongs. If the concept 'red' fails to determine a unit, there is no way in which the counting of red things can start and thus no basis for assigning any number greater than any finite number. But the basis on which Frege applies the concept of Number to a concept is that of the existence of a one–one correspondence between the extensions of concepts, and it is just not clear whether, within the framework established by Frege, there could be a one–one correspondence between the class of red things and some other class (the class of blue things, for example). This is because the distinction between concepts which do and concepts which do not determine a unit is not reflected in his formal notation; the notation suggests, but does not assert, that the concept of Number will be applicable to concepts without restriction.

Frege's universe of objects corresponds to the logical category 'object' and is the extension of the concept '$x = x$'. This means that the universe itself is an object and is thus a member of itself. More generally, for each concept there is an object in the universe which is its extension, and this is an object over and above its members, but also an object over which the concept must be defined. This again introduces a tension in Frege's treatment of classes – that between treating them as objects which are intensionally deter-mined and treating them as objects which are extensionally deter-minate.

When we give intuitive acceptance to the idea that to every

concept there corresponds a class as its extension, the picture is of a universe of objects which can be grouped in various ways to form classes. Since each object either does or does not fall under a given (well-defined) concept, it makes perfect sense to talk of the determinate totality of those which do fall under the concept. This totality will be determinate provided that the universe is. Here we do not picture the totality itself as being amongst the objects from which it may be formed. But when we look at Frege's universe of objects we find that it contains classes, classes of classes, classes of classes of classes, and so on indefinitely. For each individual object there is an infinite sequence of classes $\{a\}$, $\{\{a\}\}$, $\{\{\{a\}\}\}$, ... This means that the extension of a given concept may contain classes of an arbitrarily high complexity. From Frege's point of view this was an advantage. It was necessary if the full logicist programme for arithmetic was to be carried out because the basis of its claim was that the use of arithmetic is as universal as logic; objects of any kind, including numbers themselves, can be counted. Thus, in the expression 'the number of Fs' the concept Fx cannot be restricted to one which is defined only over individual concrete objects. None the less, Frege did insist that only *objects* could be counted; what we count is always the number of objects falling under a concept. However, this is another source of tension in his treatment of concepts. The tension is between presuming that the extensions of concepts have fully determinate membership and admitting these determinate extensions as objects on a par with all others. This latter admission appears to introduce an inevitable indeterminacy, or lack of closure into the universe of objects over which all (first-level) concepts are defined. Hence it undermines the assumption that the extensions of concepts over the universe of all possible objects have a determinate membership (are extensions to which numbers could sensibly be assigned) removing the assurance that the universe itself is determinate.

Frege exploits the fact that numbers themselves can be counted in his definition of the natural numbers. He defines $n+1$ as the class of classes having the same number of elements as the class of numbers $\leq n$, and 0 as the class of classes having the same number of elements as the extension of the concept $x \neq x$. So what we get is

$$0 = \varnothing, \ 1 = \{x : \exists R \mathscr{C}(Rx\{\varnothing\})\}, \ 2 = \{x : \exists R \mathscr{C}(Rx\{0, 1\})\}, \dots$$

The rationale for this is that it builds into the definition of the numbers an analysis of.what is involved in their application. To say that there are n Fs is to say that there is a one–one correspondence between a standard n-membered set (the number sequence up to $n - 1$) and the Fs. So

$$\text{the number of } Fs = n \equiv_{df} \text{ there are } n\, Fs \equiv_{df} \{x: Fx\} \in n$$
$$\equiv_{df} \{x: Fx\} \in \{x: \exists R\mathscr{C}(Rx\{y: y \leqslant n - 1\})\}$$

As expressed here, however, this is not a logically correct definition of the class of natural numbers, since it is an inductive definition and the legitimacy of inductive definitions needs to be demonstrated. To use an inductive definition in defining the natural numbers is, from the logicist point of view, to introduce a circularity. Frege, however, succeeded in giving a purely logical expression of the condition for an object to be an R-predecessor of an element in a set ordered by the relation R. This, together with a definition, using only logical constants, of the natural order relation for numbers, gives Frege a way of defining the set of numbers less than n. So, in the end there is nothing constructive in his definition of numbers. The numbers are not built up. Given his definition Frege is able to prove that there are infinitely many natural numbers (an actual, not a potential infinity) and the legitimacy of proof by induction over the natural numbers. But these proofs, as his definition, depend on being able to apply numbers to sets of numbers. This has the consequence that, for example {8} belongs to 1 and also that {1} belongs to 1. There is thus no sense in which Fregean classes, as extensions of concepts or as collections of objects, can be thought to be given *after* their parts. A class C may be supposed to be a determinate collection of objects, but C may be one of these objects, or some of these objects may themselves contain C, and this means that the elements need not themselves be determined or determinate independently of, or prior to, the class of which they are elements. In shifting from the Aristotelian term to the Fregean concept, with the correspondingly increased power of defining concepts from relations, the notion of 'class', as the extension of a term or a concept, has also altered. It is no longer, strictly speaking, a whole given after its parts, but nor is it a whole given prior to its parts. This is the price (if it is one) paid for seeking a univocal logical notion to replace the dual interpretation of traditional theory. Frege's logic is extensional, but his universe is a universe of all possible objects conceived as actual, and when we try to conceive this, both as a completed totality with a determinate membership and as itself an object belonging to itself, we run into contradictions.

3 Iterative Sets and Simple Types

It has been customary to blame the contradiction in Frege's system on his adoption of an unrestricted comprehension axiom – the assumption that to every first-order concept there is a corresponding object, its extension. None the less the strategy of response, with the exception of Quine's New Foundations, has been to seek to avoid the contradiction by reinstating the traditional idea that a class is a whole given after its parts (the iterative concept of set). Thus in ZF (and its extensions due to Gödel, Bernays and Morse) and in the work of Russell, limitations are placed on the comprehension axiom to ensure that neither sets nor classes can belong to themselves or to their own members. In ZF the axiom of subsets (separation) restricts the extensions of predicates to being extensions taken over an already given set, and the axiom of foundation ensures that there are no infinite descending chains under the membership relation. What is preserved is the notion of a set as something with a determinate membership and not that of a class as the totality of all possible things of a given kind.

There is an appeal here to a basic intuition about sets, sometimes called the *iterative conception of set*. There are two points to be made about this: (a) since it restricts comprehension by working from a view about sets, it is not a logicist conception and is not an explication of the notion of a class derived purely from logic; (b) it is a quasi-combinatorial, quasi-constructive notion in that it conceptualizes the universe of sets as hierarchically organized, where the levels in the hierarchy are very naturally thought of in terms of successive constructions, but they are not constructions which anyone is supposed to be able to carry out (hence the 'quasi-'). It treats sets combinatorially in that they are treated simply as collections of elements and the theory is a theory about all possible collections of given elements. Although the sets involved are not all finite, they are treated, as far as possible, by analogy with finite sets. It is worth exploring this notion at this point because it has much in common with Russell's attempt to provide a coherent logicist foundation for the notion of 'class' and with his ramified theory of types.

The simplest version of the iterative conception is provided by Russell himself with his theory of simple types. This starts from the idea that classes (or sets) are the product of divisions within a *given* universe of objects and insists that classes, as objects, are of a different type from their members and therefore do not belong to the universe from which their potential members are drawn. This reinforces the idea that a class is a whole given after its parts. But since classes are objects, they themselves can be classified and can be members of further classes. So the picture is:

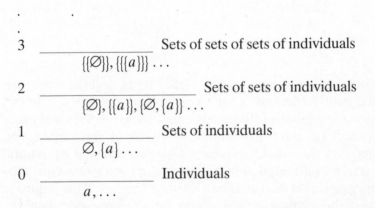

If, as in ZF, there are no individuals, then there will be no objects of type 0, just $\varnothing$ of type 1, $\{\varnothing\}$ of type 2, $\{\{\varnothing\}\}$ of type 3, and so on, i.e. there would be only one object of each type. Russell assumed that there would be a non-empty universe of individuals, but even so there would never be any infinite sets unless there were infinitely many individuals to start with. However, there can be a set of all individuals; it is an object of type 1. More generally the set of all objects of type n is itself an object of type $n + 1$.

However, it is easy to see that this rigid stratification imposes more restrictions than are required by the basic idea that classes are to be given after their members, because it allows only for classes whose members are all of a single type. It rules out, for example, $\{x: x = a \lor x = \{a\}\}$, for any given individual a, even once both a and $\{a\}$ exist. The most natural liberalization is to make the types cumulative, so that the objects of type $n + 1$ are classes of objects of type $\leqslant n$. This would give:

$$\vdots$$

3 _____ type 3
$$\{\{\varnothing\}\}, \{\varnothing\}, \{\{\varnothing\}, a\}, \{\varnothing, \{\{a\}\}\} \cdots$$

2 _____ type 2
$$\{\varnothing\}, \varnothing, \{\{a\}\}, \{\varnothing, a\}, \{\{a\}, a\} \cdots$$

1 _____ type 1
$$\varnothing, \{a\}, a, \cdots$$

0 _____ type 0
$$a, \cdots$$

So objects of type n belong to all higher types, and the type of an object is given by the least n for which it belongs to type n. This still does not allow for any infinite sets unless there are infinitely many individuals. To allow for infinite sets without this assumption requires that the whole sequence of types, indexed by natural numbers, be considered to be collectable to form a new type, type ω. This type would then itself be a set of type $\omega + 1$. If the sequence of transfinite ordinal numbers can be presupposed, then the iterative element in this conception of set allows for the formation of sets of arbitrarily high transfinite type and correspondingly for the formation of sets of arbitrarily high transfinite cardinality. For limit ordinals α we would have

$$\text{type } \alpha = \bigcup_{\alpha < \beta} \text{type } \beta$$

If this is admitted, there are two distinct ways in which sets can be 'generated' from previously given elements: (a) by forming subsets of a given set, (b) by collecting together all products of iterating the set-forming procedures applied to a given set of objects as starting point. But (b) cannot be fully generalized. It has to be restricted in such a way that the universe of all sets does not itself become a set. Here the sequence of transfinite ordinal numbers has to bear the burden of elucidating the admissible notion of iteration, and this sequence has to be conceived as absolutely non-ending, it cannot itself be allowed to form a totality (set). However, as we have seen, the transfinite ordinals cannot be developed very far into the cardinally transfinite without input from set theory itself, so from a foundational point of view, circularity threatens.

This circularity is avoided in ZF because the assumptions sufficient to provide for the ordinal number sequences are made explicit in the form of axioms. Exactly how far into the cardinally transfinite we go depends on the axioms; so that one way of extending ZF is to add higher cardinal axioms – ones which state the existence of sets of various, very high cardinalities, whose existence cannot be proved from the basic axioms (see chapter 9). But it remains the case that the basic intuition of the cumulative type hierarchy (and hence of the iterative conception of set) is incorporated within ZF in that the universe of any model of ZF can be pictured as having a cumulative type structure. This can be formally expressed in ZF itself by assigning to each set a rank (its level in the type hierarchy).

For any set x, the *rank* of x, $r(x) =_{df} \bigcup \{r(y) + 1 : y \in x\}$

$$R(\alpha + 1) =_{df} V_{\alpha+1} =_{df} P(V_\alpha) \cup V_\alpha, R(0) =_{df} V_0 =_{df} \varnothing$$

$$R(\alpha) =_{df} V_\alpha =_{df} \bigcup_{\beta < \alpha} V_\beta \qquad \text{for } \alpha \text{ a limit ordinal}$$

It can then be proved that:

1 If the highest rank of any element x is α, then $r(x) = \alpha + 1$.
2 If there is no element of x of highest rank, then there is a least upper bound on the ranks and this will be the rank of x.
3 $r(x)$ is always an ordinal number.
4 $r(\varnothing) = \varnothing$, $r(\{\varnothing\}) = \{\varnothing\}$, and for any ordinal number α, $r(\alpha) = \alpha$.

$5 \; x \in y \to r(x) < r(y)$
$6 \; x \in R(\alpha) \Rightarrow r(x) < \alpha \quad \text{and} \quad \forall x \exists \alpha (x \in R(\alpha)).$

So every set in the ZF universe appears somewhere in this hierarchy and the universe of every model of ZF can be considered as a hierarchy of this form, where the universe is the union, taken over all ordinals (which is not of course a set) of all the stages V_α. Different models are possible because neither the construction, nor the ZF axioms fully specify the membership of the power set of any given infinite set. It is this question which is central to the concern of philosophers, such as Russell, seeking a precise and consistent elucidation of the concept of 'set'.

Moreover, it is possible to prove in ZF that, for any given formula A, there are ordinal numbers α such that the restriction A^{V_α} to V_α is equivalent to A; i.e. A is true in the whole universe iff it is true in V_α. In this case V_α is said to 'reflect' A. The formal statement of this is known as the *reflection principle* – for any well-formed formula $\phi(x_1 \ldots, x_n)$ of ZF, where $x_1, \ldots, x_n$ are all its free variables,

$$\text{ZF} \vdash \forall \alpha \exists \beta > \alpha \forall x_1, \ldots \forall x_n \in V_\beta (\phi(x_1, \ldots, x_n) \equiv \phi^{V\beta}(x_1 \ldots, x_n))$$

This means that there is no way, within the language of ZF, to characterize the whole set-theoretic universe, as opposed to some member of it (some V_α). Every formula in the language of ZF, if true at all, will be true within some set which falls short of being the whole universe. So there is no way, from within ZF, of insisting that one is talking about *all* sets rather than about all sets up to a given rank.

4 Russell's Logicist Reduction

The problem from the logicist point of view is not merely that of how to introduce workable restrictions on set formation, but to determine what are the *logical* principles which govern classes interpreted as logical objects. Russell considered three possible approaches, each of which can be seen as focusing on one of three different strands of the notion of 'class'. One of these was the simple theory of types, which takes a purely extensional approach and considers classes simply as collections of objects. This does not, however, answer all the logicist's questions because logic deals in concepts, propositions and their relations. In so far as a theory of classes is to be a part of logic, it is necessary to clarify the connection between classes and concepts, and to determine either what constitutes a well-defined concept in such a way that every concept determines a class, or to provide a means of distinguishing between those concepts which do and those which do not determine classes as their extensions (between predicative and non-predicative propositional functions, in Russell's terminology).

Russell talks of propositional functions instead of concepts, or concept expressions. A propositional function is denoted by any expression obtained from a sentence which expresses a proposition by removing one or more denoting expressions – it is a function which yields propositions as values. The difference between this and a Fregean concept (which has truth values as values) is that two distinct propositional functions may express the same Fregean concept, for Fregean concepts are identical when they take the same truth values for the same arguments, whereas propositional functions are identical only when they yield the same propositions for the same arguments. The identity of a propositional function thus depends on the logical forms of the propositions which are its values. Russell calls a propositional function predicative if it determines a class. His question is how to distinguish between predicative and non-predicative propositional functions. He considered two approaches to this problem: the zig-zag theory and the ramified theory of types.

Zig-zag Theory   This starts from the assumption that predicative functions are closed under negation. If Fx is a predicative function, then so is $\neg Fx$. This amounts to assuming that every class X has a complement $\overline{X}$ which is also a class such that

$$\forall x (Fx \equiv x \in X) \,\&\, \forall x (\neg Fx \equiv x \in \overline{X}))$$

(something which is not true under any of the hierarchical approaches). If there is a propositional function Fx which is not predicative, then, for each class X, it would be that case that

$$\exists y (y \in X \,\&\, \neg Fy) \,\&\, \exists y (y \in \overline{X} \,\&\, Fy) \qquad \text{(figure 7.2)}$$

In other words, given any class X, some, but not all members of X will satisfy Fx and some, but not all, members of X will fail to satisfy Fx.

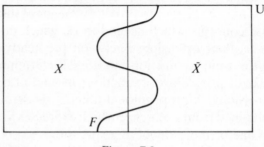

Figure 7.2

To assume that every class has a complement is to see classes as the result of dividing a universe of objects into two parts of equal standing. Russell's paradox shows that the fact that a given proposition function Fx is such that, for any object a, Fa is determinately true or false is not a sufficient condition for Fx to be regarded as creating such a division. The further requirement proposed in the zig-zag theory is that there be a 'logical symmetry' between Fx and its negation. What might this be?

It is possible to give an example of an important class of predicates of natural numbers which is closed under negation and which might therefore be predicative under Russell's proposal – the class of recursive predicates. The definition of a recursive predicate takes the form

$$Fx \quad \text{iff} \quad f(x) = 0, \quad \text{where } f(x) \text{ is a recursive function.}$$

Here the value of $f(x)$ is effectively computable. More generally it could be said that whenever there is a decision procedure for determining whether a propositional function yields a true or a false proposition, then, since a decision procedure is a procedure both for determining when $\neg Fx$ and for determining when Fx, Fx and $\neg Fx$ will be logically symmetrical predicates.

It is also possible to give examples where this is not the case:

$Fy \equiv_{df} \exists x (f(x) = y)$,   where $f(x)$ is a recursive function.

Here the procedure for computing values of $f(x)$ gives an effective means of generating a sequence of ys for which Fy holds by computing values of $f(x)$ for increasing values of x. But this does not necessarily give any information about the ys for which Fy does not hold. To know that $\neg Fy$ is true we need $\neg\exists x(f(x) = y)$ and this depends on the complete range of values of $f(x)$. It can be shown that there are propositional functions which can be defined in this way for which there can be no extensionally equivalent recursive predicate and which are thus such that there can be no definition of $\neg Fy$ in the form $\exists x(g(x) = y)$. Indeed Kleene (1967) showed that for the predicate forms

Ra	$\exists xRax$	$\forall x\exists yRaxy$	$\exists a\forall y\exists zRaxyz$	...
	$\forall xRax$	$\exists x\forall yRaxy$	$\forall a\exists y\forall zRaxyz$	...

where R is in each case general recursive, to each form with $k + 1$ quantifiers (where $k \geqslant 0$) there is a predicate expressible in the negation of the form, but not in the form itself, or in any of the forms with $\leqslant k$ quantifiers. So once quantifiers are introduced there is no automatic logical symmetry between a predicate and its negation.

An important instance occurs in connection with first-order logic itself, when it is shown that there is no effective decision procedure for a well-formed formula being a theorem. A well-formed formula A is a theorem if, and only if, there is a proof of A. This can be shown to be the case by constructing a proof. It is possible to write an algorithm generating proofs and hence theorems. But A is not a theorem only if there is no proof of A and to prove this requires showing that no proof exists. A similar example is provided by the proof of Gödel's first incompleteness theorem where he shows that the set of gödel numbers of theorems in a given formal system of arithmetic is recursively inseparable from its complement, i.e. that every recursive predicate is either satisfied by the gödel numbers of some non-theorems or is not satisfied by all the gödel numbers of theorems.

Here, surprisingly, we find a distinction which is logically similar to that made by Aristotle between kinds and attributes, but without the associated metaphysics. It is necessary to recognize a distinction between these predicates (propositional functions, or concepts) which are like decision procedures in that the predicate and its negation are 'logically homogeneous', and those predicates where this is not the case. However, it must be noted that there are no predicates which are absolutely predicative in the sense required by the zig-zag theory. The examples given are of predicate restricted to a given domain (the natural numbers). But the full definition of any recursive predicate would have to be

$$Fx \equiv_{df} x \text{ is a natural number and } f(x) = 0$$
$$\neg Fx \equiv_{df} x \text{ is not a natural number or } f(x) = 1$$

And this would yield a decision procedure only if there were a decision procedure for being a natural number. But providing an explicit definition of what it is to be a natural number is the whole problem that Frege and Russell were engaged with. Predicates and their negations can appear logically symmetrical provided that their application is restricted to an appropriate domain and they are associated with a decision procedure for determining whether any member of that domain does or does not satisfy the predicate. This is essentially what is implemented in theories of logical types. Domains, or types themselves, however, behave more like Aristotelian kinds – any predicate specifying the domain will not be symmetrical with its negation and will not in general be predicative.

This is made explicit by Wittgenstein (1914) in his criticism of Russell's theory of types. Russell insisted throughout on the logical symmetry of predicates and their negations (and from a logical point of view this is reasonable when negation is defined as a truth function). But he realized that this meant that negations could not be allowed to be absolute. He introduced logical types as the ranges of significance of propositional functions. A logical type is the domain over which a propositional function is defined and yields propositions having truth values. But then there can be no propositional function which characterizes these domains – gives a necessary and sufficient condition for membership in a logical type. Any such function Tx would have to be such that

$$\forall x (Tx \Rightarrow (\phi_n x \lor \neg \phi_n x))$$

where $\phi_n x$ ranges over propositional functions of type n. But then Tx would have to be significant both for objects of type $n - 1$ (for which it is true) and of other types (for which it is false). But then there would be objects a for which Ta makes sense (although it is false) but for which $\phi_n a$ makes no sense, and hence for which neither $\phi_n a$ nor $\neg \phi_n a$ is true. In which case $Ta \Rightarrow (\phi_n a \lor \neg \phi_n a)$ does not make sense either and can be neither true nor false.

Thus, once the means of defining concepts or propositional functions is extended beyond the traditionally accepted conjunction and negation, by the use of quantifiers, then even on a basically extensionalist view there comes to be a disparity between some predicates and their negations, a disparity which is a reflection of the difference between the two quantifiers. This difference is itself very similar to the difference between the two modalities, possibility and necessity. To prove impossibility or non-existence or necessity frequently requires moving to a different order or level, because strictly extensional methods cannot work. The zig-zag theory, by restricting predicative propositional functions to those which are logically symmetrical with their negations, would have insisted that all classes have complements which are also classes. But at best classes have complements only within some universe which is less than the universe of all possible objects.

No-class Theory and Ramified Theory Types   A crucial distinction between Russell's logicism and that of Frege is that Russell wished to revert to a purely empiricist nominalism. This is the motivation for the 'no-class' theory. Classes, as logical objects, are also logical fictions, figments of notation introduced for convenience. Class symbols do not denote independently existing objects, and classes do not belong to the universe of individual objects. Statements about classes must then be construed as statements about their members. Or, more generally, what Russell, if he is to be a thoroughgoing nominalist, requires is that any statement apparently about a propositional function must be an abbreviation for a logically complex proposition concerning values of that function (which will be propositions).

This is a reductionist requirement and one which is necessary if it is to be possible to avoid admitting any abstract objects, whether propositional functions or classes, into the ontology. In the end all meaningful propositions must be reducible to (possibly infinite) logical complexes of atomic propositions in which simple predicates or relations are applied to individuals (here quantifiers clearly have to be understood as introducing possibly infinite conjunctions and disjunctions). This does not require that logic be fully extensional because the values of propositional functions are propositions, not truth values. So it does not require that every proposition concerning a propositional function should have its truth value determined by the truth values of that function, i.e. not all statements about propositional functions have to be statements about their extensions. The constraints placed by this reducibility requirement are summed up in the vicious circle principle, which Russell claims is a logical principle. He gives various different formulations of this, one of the most succinct being:

> Whatever contains an apparent variable (a bound variable) must not be a possible value of that variable. (Russell, 1908, p. 163)

This is the logical analogue of insisting that if a set is to be determined by its members then it cannot belong to itself or contain members to which it itself belongs. So, for example, the following definition of 'n is a finite number' ($Fin(n)$)

$$Fin(n) \equiv_{df} \forall \phi (\phi(0) \ \& \ \forall x(\phi(x) \Rightarrow \phi(x+1)) \Rightarrow \phi(n))$$

would violate this condition because one of the constituent propositions of the right-hand side would be

$$Fin(0) \ \& \ \forall x(Fin(x) \Rightarrow Fin(x+1)) \Rightarrow Fin(n)$$

Since this contains '$Fin(n)$' itself it is clear that there will be an infinite regress in the attempt to reduce this proposition to its primitive constituents. The vicious circle principle rules that $Fin(n)$ cannot be a possible value of the variable ϕ. This example is significant because the above is one natural way to think of defining what it is to be a natural (finite) number, a definition which is ruled illegitimate by the vicious circle principle. The ramified type hierarchy is then generated as follows:

individuals	$a_1\, a_2\, a_3\ldots$
1st-order propositions	$Fa_1 \,\neg\, Fa_2\ \ Ga_1a_2\ \ Ha_1a_2a_3\ldots$
1st-order functions	$Fx \,\neg\, Fx\ \ Gxy\ \ Gxa_2\ \ Ha_1a_2x\ldots$
1st-order propositions	$\forall xFx\ \ \exists xFx\ \ \forall xGxa_2\ \ \forall x\exists yHa_1xy\ldots$
1st-order functions	$\forall xGxy\ \ \forall x\exists yHzxy\ldots$

Introduce $\phi!$ as a variable ranging over all first-order functions of one variable.

2nd-order functions	$\forall x\phi!x\ \ \exists x \,\neg\, \phi!x\ldots\phi!a\ \ \phi!a \Rightarrow \phi!b$
	$\ldots$
2nd-order propositions	$\forall \phi!\exists x(\phi!x \Rightarrow \phi!b)\ \ \exists \phi!\forall x\phi!x$
	$\exists \phi!(\phi!a \Rightarrow Fb)\ldots$
2nd-order functions	$\forall \phi!\exists x(\phi!x \Rightarrow \phi!y)\ \ \exists \phi!(\phi!a \Rightarrow \psi!b)\ldots$

Introduce new variables to range over second-order functions of one free variable, and repeat.

A *logical type* is the range of a free or a bound variable and variables range only over individuals or predicative functions.

The *order* of a propositional function is the next order above that of the highest order variable it contains, bound or free, i.e. it is $n + 1$ if all its variables, bound or free, are of order $\leqslant n$.

A *predicative propositional function* is a propositional function containing one free variable whose order is next above that of its free variable (figure 7.3). Here 3, 1, for example, indicates propositional

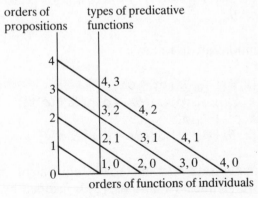

Figure 7.3

functions of order 3 with a single free variable of order 1. Only functions containing one free variable are mapped. All functions and propositions on the same diagonal have the same order. Thus functions of a given order are logically heterogeneous. On the other hand it should be noted that there is no variable which can range over all the propositional functions of a given order, nor is there any variable which can range over all propositional functions of individuals, because these functions may be of arbitrarily high order. It is therefore impossible to quantify over all such functions. It is only possible to quantify over predicative propositional functions and the predicative functions of a given order then form a type. The type of a predicative function is numerically the same as its order and is the next above the type, or order, of its free variable. Predicative functions are those which are viewed as having classes as their extensions, or which, more strictly, have to do the work of classes in that they make it possible to introduce class symbols as contextually defined symbols whose use can always be eliminated in favour of reference to predicative propositional functions.

Russell's strategy here is the same as that which he used in his theory of definite descriptions. It is to define class symbols as incomplete symbols, which have no meaning in isolation (and thus do not stand for any object). They have a meaning only when used in the context of a sentence. The contexts which require definitions are:

$$\{x: \theta x\} = \{x: \phi x\} \quad F(\{x: \theta x\}) \quad y \in \{x: \phi!x\} \quad y \in \{x: \theta x\}$$
$$\{x: \phi x\} \in \{x: \theta x\}$$

and the definitions are as follows:

$$\{x: \theta x\} = \{x: \phi x\} \equiv_{df} \forall x(\theta x \equiv \phi x) \tag{1}$$
$$F(\{x: \theta x\}) \equiv_{df} \exists \phi! \forall x((\theta x \equiv \phi!x) \,\&\, F(\phi!x)) \tag{2}$$
$$y \in \{x: \phi!x\} \equiv_{df} \phi!y \tag{3}$$
$$y \in \{x: \theta x\} \equiv_{df} \exists \phi! \forall x((\theta x \equiv \phi!x) \,\&\, \phi!y) \tag{4}$$

Here (1) simply says that materially equivalent propositional functions have identical extensions. (2) is justified by the thought that any statement about the extension of a function θx should be equally true (or false) of the extension of any function materially equivalent to it and that a higher order function only determines a class if there is some predicative function to which it is materially equivalent. (3) is what one would expect if $\{x: \phi!x\}$ is to be the extension of $\phi!x$. (4) follows from (2) and (3), which can similarly be used to give a reading of $\{x: \phi x\} \in \{x: \theta x\}$.

The question which then becomes pressing from the point of view of being able to express arithmetic within the logical framework of the ramified theory of types is whether there are 'enough' predicative functions of numbers to represent every possible subset of the natural numbers. Might it be the case that there are some sets of numbers which can only be defined as the extensions of higher order propositional functions, i.e. for which there is no extensionally equivalent predicative function? If this were the case, then quantification over the predicative functions defined over the natural numbers could not be a substitute for quantifying over all subsets of natural numbers, and the class of predicative functions will not do duty for the power set of the natural numbers. But if every higher order function were extensionally equivalent to some predicative function, then quantification over predicative functions would serve as a substitute for quantification over all subsets of the natural numbers. The assertion that every higher order function is extensionally equivalent to a predicative function is Russell's *axiom of reducibility*:

For any propositional function Fx, $\exists \phi! \forall x (Fx \equiv \phi!x)$

This would, for example, legitimize the definition

$$Fin(n) \equiv_{df} \forall \phi!((\phi!0 \ \& \ \forall x(\phi!x \Rightarrow \phi!(x + 1))) \Rightarrow \phi!n)$$

as a characterization of the natural numbers. $Fin(n)$ would be a higher order function and not a possible value of the variable $\phi!x$. But $\phi!x$ would range over all possible sets (classes) of numbers. If the axiom of reducibility were not assumed, then the statement of the principle of induction as a single principle would be impossible because there can be no quantification over the higher order functions taking numbers as arguments. Similarly the assertion of the identity of indiscernibles

$$\forall \phi !(\phi !x \equiv \phi !y) \Rightarrow x = y$$

is only possible if the axiom of reducibility holds. Without it one cannot assume that if x and y agree in all first-order properties, they will agree in all higher order ones.

But are there any grounds, consistent with the philosophic position which motivates the construction of the ramified hierarchy, for supposing that the axiom of reducibility holds? The axiom asserts, for any given non-predicative propositional function, the existence of an extensionally equivalent predicative function. If it were assumed at the outset that every propositional function of the ramified hierarchy determines a class as its extension, then the axiom of reducibility could be proved as follows:

Suppose that for any propositional function ϕ, of whatever order

$$\exists \alpha \forall x (\phi x \equiv x \in \alpha)$$

Then $x \in \alpha$ is a predicative function extensionally equivalent to ϕx.

But the converse is not true, for the axiom of reducibility asserts the existence of a propositional function, not of a class, and the criteria of identity for classes and propositional functions are not the same, unless the logic and/or language is presumed to be fully extensional, i.e. unless all higher order functions $F\phi$ are such that

$$\forall x (\phi !x \equiv \theta !x) \equiv (F\phi ! \equiv F\theta !)$$

If this were assumed, then the difference between Russell's propositional functions and Frege's concepts would be eliminated, as extensionally equivalent propositional functions would be identical, under the identity of indiscernibles, which is the definition of identity used by Russell. In this case, if

$$\exists \phi !\forall x (\theta x \equiv \phi !x)$$

there would be at most one such $\phi !x$ and it could then, for all intents and purposes be treated as the function $x \in \{ y : \theta y \}$.

Now, as has already been noted, if attention is restricted to arithmetical contexts, and if the logicist claim, that all arithmetical notions are logically definable (by use of quantifiers and identity alone), is correct, then all the higher order functions used in arithmetic will be extensional. But it is also part of the logicist claim that arithmetic is not special; if it is reducible to logic, it is not supposed to be reducible to a logic which is the logic of arithmetic, or even of mathematics, but to logic as a universal canon of reasoning. Here it is not reasonable to assume that all higher order propositional functions are extensional because in normal discourse we say things about propositional functions which depend on their meaning, on the way they are defined (their semantic characteristics) as well as about their extensions. In addition we express attitudes, such as belief or disbelief toward the content of propositions, which are not based on a knowledge of their truth values.

Whereas expressions of personal attitudes toward propositions might plausibly be ruled to be extrinsic to the identity of propositions and propositional functions, and thus be excluded from the higher order contexts which determine the identity of propositional functions via the identity of indiscernibles, the same cannot be said of statements about semantic characteristics. The problem is that the decision here rests precisely on how the identity of propositions and propositional functions is to be conceived. The immediate difficulty could be circumvented in the manner of Quine (1953, ch. VIII). He insists that all this apparently higher order non-extensional talk is generated by confusing the *use* of a linguistic expression with its *mention*. The thoroughgoing nominalist should no more countenance propositions and propositional functions as distinct from the linguistic symbols which express them, than he should countenance any other abstract entities. All apparently higher order, non-extensional talk of propositions and propositional functions is really metalinguistic talk of sentences and sentential functions – talk about linguistic expressions which are objects in their own right. All higher order extensional talk is really about classes, or sets, and therefore also about things which are objects in their own right. So the need for higher order logic disappears and the hierarchy of propositional functions is replaced by a hierarchy of first-order languages.

However, it can readily be seen that this solution constitutes a capitulation on the point of trying to carry out the logicist programme. It forgoes the hope of reducing set theory to logic. In *Set Theory and Its Logic* Quine (1963) very carefully shows just how far he can go with a theory of virtual classes. It stops at the point where quantification over classes is required. But from the point of view of the Fregean reduction of logic to arithmetic, this is really rather early because higher order quantification was needed to define 'the number of *F*s is the same as the number of *G*s' and 'there is a one–one correspondence between the *F*s and the *G*s'.

In addition, linguistic expressions, as the objects of metatheory, are also abstract objects whose status is not inherently clearer than that of numbers. Quantification over the well-formed formulae of a formal language is quantification over an infinite domain of formula types, not over actual formula tokens, since there is no assurance that there ever will exist token instances of all possible formula types. So on foundational grounds, for a nominalist, there is little real gain in shifting from a hierarchy of logical orders to a hierarchy of orders of languages each of which is itself a first-order language. There is a gain from the point of view of being able to insist that all logic is extensional, but it is not a way of independently justifying that view.

Given that Russell does not adopt Quine's approach, but persists in confusing use with mention, he can justifiably claim that the axiom of reducibility is weaker than the unrestricted comprehension axiom and that it does not amount to asserting the independent existence of classes. But he cannot get round the fact that the axiom of reducibility is an existential assumption. Russell's predicative functions of individuals are all either propositional functions derived from atomic propositions by removal of an individual constant or from truth functions of atomic propositions (including among truth functions the possibly infinite conjunctions and disjunctions expressed by means of quantification over individuals). So exactly what predicative functions there are is very much dependent on the given stock of atomic propositions. They could be considered to consist of expressions of the characteristics which an individual can have in its own right, or of the relations in which an individual can stand to a finite number of other individuals, or to every other individual. Non-predicative functions of individuals are those which characterize an individual by reference to the totality of its first (or higher) order characteristics. The axiom of reducibility then asserts that in principle, for the purposes of classification, quantification over propositional functions is eliminable; i.e. that all classification of individuals could, in principle, be based on their intrinsic characteristics or on their first-order relations to other individuals. This is reflected in the form of the principle of identity of indiscernibles which results from the axiom of reducibility – individuals are identical if they agree in all their predicative properties. In other words, the higher order functions are also really just truth functions of atomic propositions.

Now this claim is wholly consistent with the form of nominalism which gave rise to the ramified theory of types. The problem, from the point of view of an independent justification of the logicist claims about mathematics, is that it is an assertion of a metaphysical position and not a philosophically neutral claim. Moreover, when consistently asserted in this way, the axiom of reducibility will not be sufficient to provide a foundation for mathematics. The metaphysical position could be maintained in a legislative form, if it were supposed that there are independent criteria for determining the totality of atomic propositions. (It would be on a par with claims made by methodological individualists about the reducibility of all apparently societal facts and sociological classifications of individuals to propositions about individuals which contain no reference to society as a whole.) The legitimate theory of classes would then be that which could be constructed on this basis.

Note that the restriction here does not derive from qualms about the actual infinity, for the actual infinite has already been embraced with the idea of infinite truth functions, nor would they be based in conceptions of effective decidability. Even so the restrictions imposed would mean that the resulting theory of classes would not suffice for anything much in the way of transfinite arithmetic. If there are an infinite number of individuals and a denumerable infinity of atomic propositions, there can still only be denumerably many predicative functions of individuals, since all definitions of such functions have to be of finite length. This would mean that there would not be non-denumerably many subclasses of a denumerably infinite class of individuals. This strong reductivist position would then have a similar effect to the denial of the actual infinite (in that there would be no legitimation of transfinite cardinal arithmetic) even though this was not its rationale.

However, Russell did not adopt this position, which in all consistency he should have adopted (and which Wittgenstein urged). Instead he assumed that the axiom of reducibility gives him, in effect, all the classes he needs for classical mathematics, and thus that it gives him non-denumerably many subclasses of an infinite class of individuals. But in making this assumption Russell has either effectively smuggled in the theory of sets to underwrite the possibility of non-denumerably many truth functions over his atomic propositions, or he has gone back on his antirealist attitude toward propositional functions. Either way his thoroughgoing nominalism has been abandoned.

Once the strongly nominalist stance is abandoned, and a realist attitude toward sets or propositional functions is adopted, there is no reason to believe that the vicious circle principle is a logical principle and the ramified type hierarchy collapses into the simple theory of types. If sets and propositional functions exist there is nothing wrong with impredicative definitions. Since a definition merely has to secure unique reference for a linguistic expression, it is not required to be a principle according to which that reference could be constructed. That the definition of a linguistic expression contains an apparent variable ranging over that expression does not automatically mean that the entity to which the expression refers embodies a circularity.

So, although the axiom of reducibility could be justified in a manner consistent with the philosophic position which leads to the vicious circle principle and the ramified theory of types (the thoroughgoing nominalism which treats all abstract objects as constructed fictions), this would not yield an axiom sufficiently powerful to ensure the existence of the classes required for the development of classical mathematics. But when the axiom is adopted in the form required for the development of classical mathematics, it involves a realist commitment to propositional functions (and/or their extensions) which undercuts the philosophic position motivating the ramified hierarchy. In either case Russell's logicist reduction fails. The development of classical mathematics apparently requires a commitment to the existence of classes (and/or propositional functions) which goes beyond anything assured by (definable within) Russellian logic alone.

However, the issue here is not just one of realism or anti-realism with respect to classes. It is very much an issue concerning the notion of 'class' itself. Russell's ramified type hierarchy represents an attempt at a rigorous development of the notion of 'class' as a purely logical notion (the extension of a propositional function) which none the less seeks to treat classes extensionally as being not only given after their members but also as being no more than a collective way of referring to their members. Russell's classes are collections only of actual objects and his universe of individuals is restricted to the empirically actual. But even Russell's extensionalism does not make it possible to identify the logical notion 'class' (extension of a linguistically expressible propositional function) with the mathematical notion 'set' as simply a collection of objects, not necessarily having anything in common and not necessarily having any principle of collection. It is this opposition which is at the heart of misgivings about the axiom of choice (which Russell also had to adopt, in the form of the multiplicative axiom, in order to develop an arithmetic of transfinite cardinal numbers). Does the existence of a choice set require there to be a principle of choice, or is set existence wholly independent of defining principles, principles which would allow a (possibly infinite) being to make the selection?

The logicist claim that arithmetic is nothing but logic was based on the universal applicability of arithmetic, its topic neutrality – a feature which it shares with logic. Now it is a disadvantage of Russell's hierarchies, in contradistinction to Frege's untyped domain of objects, that his definitions of numbers, his statements of the principle of numerical induction, etc. have to be iterated at every level of the hierarchy. This can easily be done, but it means that the logical formulation at any level then fails to capture what was perceived to be important about arithmetic. Arithmetical laws, like logical laws, hold for any and every level of the hierarchy (in Russell's terminology they are 'typically ambiguous') and yet they cannot, with logical propriety, be stated in such a way that this can be revealed. This represents another respect in which Russell's claims to have provided an adequate reduction of mathematics to logic by means of the ramified type hierarchy must be rejected.

Russell did not provide a foundation for pure mathematics so much as an account of how mathematics is applied, given a particular vision of the nature of the world to which it is to be applied. The status of arithmetical laws, and their universal applicability in particular, remains unelucidated because it remains part of the presupposed background within which the theory is elaborated.

If Frege was right in thinking that the characteristic feature of arithmetic (and along with it algebra, arithmetized analysis, set theory, etc.) is that it is topic neutral and universally applicable, it would not necessarily follow that mathematics is reducible to logic. It would, however, follow that when considering possible numbers, functions, sets, etc. one cannot restrict attention only to those definable or constructible by the means suggested by any particular application or within any particular notation. Its pure theory must concern all possible entities of the kind without prejudice as to how these might be realized. But the price to be paid for maximum applicability has to be extensional indefiniteness.

8

Independence Results and the Universe of Sets

The proofs (mentioned at the end of chapter 6) of the independence of the generalized continuum hypothesis (GCH) and the axiom of choice (AC) from the remaining ZF axioms reveal the extent to which the ZF axioms fail to give an extensionally determinate specification of the universe of sets. By looking at the strategies employed in providing these results it becomes apparent that this is not just a failure to determine the extension of a well-defined concept, but fully to determine the concept of *set* itself. In particular the ZF axioms do not force a decision on the question of whether, or how, the notion of set is to be tied to its logical roots as the extension of a term, or allowed the full freedom of the notion of an arbitrary collection, suggested by its combinatorial origins.

1   Gödel's Constructible Universe

Gödel proved that if ZF is consistent then so is ZF + GCH + AC, and he did this by showing that there is a model of ZF which consists only of sets which, in the sense of Russell's ramified type theory, have been predicatively defined. This universe of sets (the universe of constructible sets) is, however, not identical with that consisting of the extensions of Russell's predicatively defined propositional functions. For the notion of a constructible set, whilst being logically constrained, is not a pure logical notion. The constructible sets are those which are definable in ZF by expressions of the language of ZF which quantify only over sets which have been previously defined. Whereas the ramified type hierarchy assumes iteration only up to ω, Gödel's construction is allowed to be iterated up to any transfinite ordinal. The constructible universe is thus generated by taking the sequence of transfinite ordinal numbers (and the legitimacy of definition by transfinite induction over it) for granted, without imposing any condition of predicative definability on the ordinals themselves. But if the universe of constructible sets outruns that of Russell's predicative propositional functions in this respect, it is much more narrowly confined in another – it is restricted to the sets definable within ZF. This means (a) that, since there is no universe of individuals, there are no sets other than those ultimately constructible from the empty set, and (b) that only 'structural' characteristics of sets (characteristics expressible in terms of the single non-logical primitive '∈' and the first order quantifiers), or relations based on these, can be used to characterize sets and collect them into further sets. The universe of constructible sets is thus very much a mathematician's, rather than a logician's, universe.

The constructible sets, then, are those definable in ZF by expressions which quantify only over sets which have been previously defined. But the subset and replacement axioms of ZF impose no such restriction. They allow quantification over the whole universe of sets. The following would, for example, be a subset of ω definable in ZF, whose existence can be proved by reference to the axiom of subsets, but which is not constructively defined:

$S =_{df} \{x: x \in \omega \ \& \ Px\}$, where Px if and only if there is a partition of ω into x disjoint sets, none of which contains arithmetic progressions of arbitrary length.

To consider whether $n \in S$ requires considering *all* partitions of ω into n disjoint sets. Because the existence of such impredicatively defined sets can be proved in ZF, it is a non-trivial matter to show that there is a model of ZF which satisfies the predicativity constraint.

As has been said, Gödel obtained his model by allowing predicative contruction of sets to be iterated up to any transfinite ordinal. Of it Gödel says:

This model, roughly speaking, consists of all 'mathematically constructible' sets, where the term 'constructible' is to be understood in the semi-intuitionistic sense which excludes impredicative procedures. This means 'constructible' sets are defined to be those sets which can be obtained by Russell's ramified hierarchy of types, if extended to include transfinite orders. The extension to transfinite orders has the consequence that the model satisfies the impredicative axioms of set theory because an axiom of reducibility can be proved for sufficiently high orders. (1938, p. 556)

He adds that, in particular, GCH follows from the fact that all constructible sets of integers are of order less than ω_1, all constructible sets of sets of integers of order less than ω_2, and so on. This means that it can be shown that, for the set S in the above example, Px is extensionally equivalent to a condition expressed by quantifying only over constructible sets of order less than ω_1.

The constructible universe is defined in the same manner as ranks were defined on p. 157. If x is a set, put

$$y \in Def(x) \equiv_{df} y = \{z : z \in x \ \& \ A_x(x, t_1, t_2, \ldots, t_n)\}$$

where $A_x(x, t_1, t_2, \ldots, t_n)$ is a well-formed formula of ZF and A_x is A with all bound variables restricted to x, and all its terms $t_1, t_2, \ldots, t_n$ are members of x.

Put

$$L_0 = \varnothing$$

and for all ordinals $\alpha > 0$ put

$$L_{\alpha+1} = Def(L_\alpha) \cup L_\alpha$$

$$L_\alpha = \bigcup_{\alpha < \beta} L_\beta \quad \text{if } \alpha \text{ is a limit ordinal.}$$

A set x is then *constructible* if, and only if, there is an ordinal α such that $x \in L_\alpha$.

If L is the class of all constructible sets and V is the class of all sets, then V = L, or

$$\forall x \exists \alpha (x \in L_\alpha)$$

says that every set is constructible.

It can then be proved that if ZF has a model (is consistent), so that there is a sequence of ordinals available, then ZF + V = L also has a model (is consistent). In other words, the universe of constructible sets forms a model of ZF providing ZF has a model. It can then be shown that in this model AC and GCH hold, i.e.

$$ZF + V = L \vdash AC + GCH$$

That AC holds in L is shown by showing that every constructible set can be well ordered. That this well-ordering is itself constructible, and that the assumption that every set can be well ordered entails the axiom of choice. Every constructible set can be well ordered because the universe L is constructed from $\varnothing$ in a well-ordered sequence of stages. Each stage after L_0 is the result either of adding to a previous stage all sets constructible from it, or of collecting together previous stages. For any ordinal number α, $L_{\alpha+1}$ is a set which can be well ordered if L_α can. Since the sets added to L_α are all definable by a formula in the language of ZF of the form $A(x, t_1, t_2, \ldots, t_n)$ an ordering on them can be derived from an 'alphabetical' ordering of the denumerably many formulae $A(x, y_1, y_2, \ldots, y_n)$ together with the well-ordering of L_α, since the terms t_i are restricted to standing for members of L_α. So if L_α is well ordered and the sets constructible from it, and not already belonging to it, are added onto the end of its ordering in their order, $L_{\alpha+1}$ will be well-ordered. If there are well-orderings of L_β for all $\beta < \alpha$, L_α is well ordered as follows:

for x and y in L_α suppose β_1 and β_2 are the least ordinals such that $x \in L_{\beta_1}$ and $y \in L_{\beta_0}$ respectively. Put $x < y$ if, and only if, $\beta_1 < \beta_2$, or $\beta_1 = \beta_2$ and x comes before y in the ordering of β_2.

The definition of this ordering can then be extended to cover the whole universe. Thus the whole constructible universe can be well ordered and so can any set S within it, since all members of S will be members of the constructible universe. If every set can be well ordered, there is a principle available for selecting one element from every member of any given set of mutually disjoint sets – well order each of these mutually disjoint sets and select the least element from each. However, to show that AC holds in the universe of constructible sets it must additionally be shown that, for any given constructible set of mutually disjoint sets, there is a *constructible* set which represents a choice function for it. This can be done by showing that orderings, similar to those mentioned above, of each L_α can be predicatively defined by ZF formulae.

At each stage of the construction of the universe L, $L_{\alpha+1}$ is denumerable if L_α is, thus the only way in which non-denumerable constructible sets can be obtained is by iterating the construction process non-denumerably many times. This means that L_{ω_1} (the union of ω_1 denumerable sets) will be of cardinality $\aleph_1$. So if all the constructible subsets of ω are already present in L_{ω_1} (i.e. no new subsets of ω can be defined by quantifying over levels higher than this), the cardinality of the constructible power set of ω must be $\leqslant \aleph_1$. But if L is to be a model of ZF, the cardinality of the power set of ω in L cannot be $\aleph_0$ so it must be $\aleph_1$. Again, in order to complete the proof that $2^{\aleph_0} = \aleph_1$ is true in L it is necessary to show that the above facts entail that there is a constructible set of ordered pairs which is a one–one correspondence between $2^{\aleph_0}$ and $\aleph_1$, or between subsets of ω and members of ω_1.

2 Cardinals and Ordinals in Models

This distinction between seeing what we might be able to say *of* L (or of the universe of any other proposed model of ZF) and what is true *in* L is very necessary. The most dramatic demonstration of the necessity for this distinction arises when considering the consequences for ZF of the Löwenheim–Skolem theorem. The Löwenheim–Skolem theorem says that if any first-order theory T has an infinite model, then it has a model which is denumerably infinite. Since ZF is a first-order theory which, because it contains an axiom of infinity, has only infinite models, it too must have a denumerable model. But how can ZF, with the power set axiom, in addition to the axiom of infinity, have a denumerable model? Surely the power set axiom, by requiring the existence of the set of all subsets of ω, requires the existence of a non-denumerable set.

However, for M to be a model of ZF all that is required is that there does not exist *in M* a set of ordered pairs which is a one–one correspondence between ω and 2^ω, or between ω and $P\omega$. In other words, since the condition for an *in*equality of cardinalities is the *non*-existence of a relevant one–one correspondence, a rather sparse universe of sets is able to make sets look unequal in cardinality when, from the point of view of a larger universe, they would look equal in cardinality. Thus it is possible to have a model M for ZF whose universe is denumerable (i.e. there is a one–one correspondence between it and the natural numbers) but which contains many sets which are non-denumerable from the point of view of M, because M does not contain one–one correspondences between these sets and the natural numbers.

Similarly, the universe of M may well be a set, but it cannot be a set *in M*; i.e. it cannot belong to itself and, if M is to be a model of ZF, it must be a set which cannot be formed by applying any of the ZF axioms to the sets in M. Likewise the ordinal numbers in M, from the point of view of M, cannot appear to contain a largest, nor can they form a set in spite of the fact that if M is a denumerable model it can contain only denumerably many ordinal numbers. Again, this means that if, from outside M, it is possible to put a least upper bound on the ordinal numbers in M, this will have to be done by an ordinal number which cannot be 'reached from below' by applying definable functions to ordinals in M.

But M need not itself contain any ordinals, apart from ω, which are of this kind. An *accessible cardinal number* is an initial ordinal number which can be 'reached from below'; that is, it can be reached either by adding together a smaller number of smaller ordinals, or be expressed as a value of one of the functions 2^α or $2^{\aleph_\alpha}$ for some smaller ordinal α. ω is the first ordinal which cannot be reached in this way, but the axiom of infinity assures the existence of $\omega (= \aleph_0)$ in all models of ZF. Thus an *inaccessible cardinal number* is defined as being a cardinal number greater than ω and which is not accessible. It can then be proved (Drake, 1974, pp. 109–110) that if μ is an inaccessible cardinal, then the sets in the cumulative type hierarchy of rank $\leq \mu$, i.e. the sets in V_μ, form a model for ZF + AC. If it were possible to prove in ZF the existence of an inaccessible cardinal, it would then be possible to prove in ZF that ZF has a model, since V_μ would be a set whose existence could be proved. But this would mean that ZF was a formal system powerful enough to be able to prove its own consistency. By Gödel's second incompleteness theorem we know that no consistent formal system capable of expressing arithmetic can prove its own consistency. Therefore, if ZF is consistent there must be models of ZF which do not contain any inaccessible cardinal numbers. If there are inaccessible cardinals and α is the least such, then V_α would be a model of ZF + AC in which all the cardinals in the model were accessible. If the existence of inaccessible cardinals is added to ZF + AC as an additional axiom then

$$\text{ZF} + \text{AC} + \text{axiom of inaccessibles} \vdash \text{there is a model of ZF} + \text{AC}$$

One can then go on to define *hyper-inaccessible cardinals* as inaccessible cardinals μ which have μ inaccessible cardinals below them. It can then be shown that if μ_0 is the first hyper-accessible cardinal, V_{μ_0} is a model of ZF + AC + axiom of inaccessible cardinals + there are no hyper-inaccessible cardinals. Then if we say that a cardinal μ is *hyper-hyper-inaccessible* if there are μ hyper-inaccessible cardinals below μ, and μ_1 is the least hyper-hyper-inaccessible cardinal, then V_{μ_1} is a model of

$$\text{ZF} + \text{AC} + \forall\alpha\exists\beta(\beta > \alpha \ \& \ \beta \text{ is hyper-inaccessible}) + \text{there}$$
are no hyper-hyper-inaccessible cardinals

And so on. Thus the addition of successive axioms asserting the existence of 'large' cardinals successively strengthens the system. Basically a set of axioms asserting set existence gives us a way of delimiting, from outside the system, a class of cardinals – all those required to exist by (or which can be proved to exist by reference to) these axioms. Application of Cantor's third principle of generation for transfinite ordinal numbers, the principle of limitation (see p. 106), then makes it possible to define (and so postulate the existence of) a number which is the first number after all of these, which will be the first number of a new kind.

Attention here will be restricted to standard transitive models of ZF (for an alternative route to independence results using non-standard, Boolean valued models, see Bell (1977)). A *standard transitive model of ZF* (a) is really made up of sets, so that '$\in$' is given its normal reading of 'belongs to' or 'is a member of' as applied to sets, and (b) is such that if x is a set in the universe of the model, then all members of x are also sets in the universe of the model. In such models ordinal numbers are not subject to the same sort of relativity as cardinal numbers, the ordinals are thus said to be *absolute* in such models. This is because an ordinal number just *is* the set of all ordinal numbers less than itself. So if an ordinal number α belongs to a transitive model, all the ordinal numbers less than α also belong to that model. Moreover, the order of the ordinal numbers is determined by the membership relation, since $\alpha < \beta$ if, and only if, $\alpha \in \beta$. So on standard interpretations of '$\in$', the ordering of the ordinals is fixed independently of what other sets the universe may or may not contain. Thus the only difference there can be in standard models as regards the ordinals is that one model may extend further up the ordinal number sequence than another. Here it should be remembered that once one gets beyond ω, the question of how far one can go on 'counting' into the transfinite does not depend merely on repetition of the operation of adding one, or of taking the limit of an infinite sequence of additions, but also, crucially, on the ability to collect all previously given ordinals into a set so that addition and other definable 'arithmetic' operations can begin all over again. The question of what initial ordinals there are is thus not settled by the definition of the ordinal numbers alone but also depends on what sets of ordinal numbers there are.

The fact that both the existence of a well-ordering of a given set and the establishment of a relation of cardinal equality between two given sets requires the existence of further sets (of ordered pairs of elements of the given sets) suggests a strategy for constructing models of ZF in which AC and GCH fail. This does not involve making the power set of ω too large to be well ordered or to be put into one–one correspondence with ω_1, but in making the universe of the models small enough not to contain any well-ordering of $P\omega$, on the one hand, or not to contain any one–one correspondence between $2^{\aleph_0}$ and $\aleph_1$, on the other. Indeed Cohen (1966, pp. 108–9) shows that for any axiom system containing ZF and which is consistent with $V = L$, one cannot prove the existence of a non-denumerable standard, transitive model in which AC is true but CH is false. This, intuitively, is because in any non-denumerable model in which AC holds there will have to be non-denumerable ordinals. If a standard model contains a non-denumerable ordinal it must (in the light of the absoluteness of the ordinals in standard transitive models) also contain all denumerable ordinals. But every real number is constructible from a denumerable ordinal, so in this model every real number will be constructible, and if every real number is constructible CH is also true.

However, the construction of a denumerable model in which either AC or GCH fail is not a straightforward matter because the constructible universe is, in a certain sense, the smallest universe of sets we can have. Every other model will have to contain a submodel consisting only of constructible sets and AC and GCH both hold in a universe consisting only of constructible sets. By reflecting on the methods used by Gödel it is possible to show that the same methods cannot be used to provide a model in which either AC or GCH are false.

3 Inner Models

The method used by Gödel is an example of what has subsequently been called the method of 'inner models' (Shepherdson, 1951). An *inner model* is given by defining a class using a well-formed formula of ZF containing one free variable (such as $\exists \alpha (x \in L_\alpha)$) which is such that for each axiom A of ZF the relativization of A to this class is provable in ZF. So, for example, in proving that the constructible sets form a model of ZF, one shows that the power set axiom, when relativized to constructible sets, can be proved in ZF. The *relativization* of a well-formed formula F to a class C is the well-formed formula obtained from F by restricting all quantifiers in F to C and requiring all terms in F to stand for sets belonging to C. So the required relativization of the power set axiom to the constructible universe is

$$\forall x \exists \alpha (x \in L_\alpha \Rightarrow \exists y \exists \beta (y \in L_\beta \ \& \ y = P^L(x))$$

where $P^L(x)$ is the power set in L of x (the set of constructible subsets of) thus

$$\forall z (z \in P^L(x) \equiv \exists \alpha (z \in L_\alpha \ \& \ y \in z \Rightarrow y \in x))$$

If a class C of sets can be defined by a formula of ZF and it can be shown that each of the axioms of ZF remains a theorem of ZF when it is relativized to C, then it must be the case that in any model M of ZF there must be a submodel M^C consisting of those sets in M which belongs to the class C. Since the class L, of constructible sets, satisfies this condition, it follows that every model M of ZF contains a constructible inner model M^L.

But it can also be shown that there is a model M of ZF which is minimal in the sense that every inner model of M coincides with M. This model consists of the '*strongly constructible sets*'. These are the sets which are not merely definable in the language of ZF, but which can also be proved to exist.

$$M_0 = \varnothing$$
$$M_{\alpha+1} = M_\alpha \cup Prov(M_\alpha)$$
$$M_\alpha = \underset{\alpha<\beta}{\cup}\, Prov(M_\beta)\ \text{for}\ \alpha\ \text{a limit ordinal}$$

where $Prov(x)$ consists of all sets of the following forms, with all definitions relativized to x:

1 $\{y, z\}$ for $y, z \in x$
2 the sum set of y, if $y \in x$
3 $0, 1, 2, \ldots$
4 the set of all y in x such that $z \supseteq y$, for a given $z \in x$
5 if f is a single valued function in x defined by a formula relativized to x using as constants sets in x, then if $a \in x$, $\{z : \exists y(y \in a\ \&\ fy = z)$ is in $Prov(x)$

A set x is *strongly constructible* (belongs to the class M of strongly constructible sets) if, and only if, $\exists\alpha(x \in M_\alpha)$.

Each of 1–5 corresponds to an axiom of ZF. All sets generated in this way will belong to L, but there is a crucial difference. In $Def(x)$, used in the construction of L, $x = \{z : z \in x\ \&\ z = z\}$ always belongs to $Def(x)$, since '$z = z$' is a well-formed formula of the language of ZF. Thus, whatever stage has been reached in the constructible universe, L, it is always possible to collect *all* the sets formed at that stage into a further set. This means that there can be no end to the process of construction; each stage brings essentially new sets. But in the case of *Prov*(*x*), used in the definition of M, we do not automatically get $x \in Prov(x)$, since item 4 only adds the set of all subsets in *x* of a given *member* of *x*, not of *x* itself. This corresponds to the limitation that is built into the ZF axioms in order to prevent the universe of sets being itself a set. It can be shown that the strongly contructible sets form an inner model of ZF and that there is an α such that M_α is a model of ZF. Hence there must be at least such α, say α_0. In the light of the Löwenheim–Skolem theorem, the universe of this model must be denumerable, i.e. M_{α_0} cannot contain any non-denumerable ordinals.

The strongly constructible sets, M, can be shown to constitute a minimal model for ZF in the sense, given above, that every inner model of M coincides with M. This means not only that V = L holds in M, but also that the sets in M also satisfy any other condition *Q* which, by means available in ZF, can be used to define an inner model. Thus for each such condition $Q, V = L \& \forall x Q x$ will hold in M. It follows that one cannot hope to define an inner model defined by a condition *Q* such that in ZF one could prove that $\exists x (Qx \& \neg \exists \alpha (x \in L_\alpha))$, for if one could, V = L would fail in every model satisfying $\forall x Q x$. But in the minimal model, M, both $\forall x Q x$ and V = L are true. Since $ZF + V = L \vdash AC + GCH$, it is also impossible to provide an inner model in which either AC or GCH fail, i.e. for which one could prove

$$ZF \vdash \neg GCH^Q \quad \text{or} \quad ZF \vdash \neg AC^Q.$$

4 Generic Sets

The above limitations on the method of inner models required the adoption of a different strategy for showing the independence of AC and GCH from ZF. In particular it was necessary to construct models which contained sets other than the constructible sets but which were still denumerable models. Cohen (1966) devised a method of extending the minimal model M_{α_0} to give larger models (figure 8.1). This involved adding new 'generic' sets and then taking all sets constructible from these and the sets in M_{α_0}. The first point at which it is possible to introduce sets which are not constructible is when considering infinite subsets of ω, since all finite sets, together with ω itself, are constructible.

The intuition which suggests that there is no reason to suppose that every set is definable by a condition which all its members must satisfy (that there is no reason to suppose that all sets are classes) is the intuition that a set is *any* collection of previously given objects. Thus the power set of ω should include all possible collections of elements of ω whether there is a defining condition or not. One way

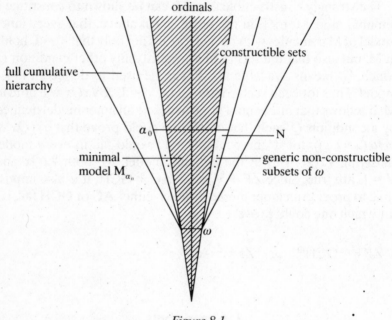

Figure 8.1

of getting a 'bigger' model of ZF is thus to introduce infinite subsets a_δ of ω which are conceived as being specifiable only by a denumerably infinite list of conditions of the form '$n \in a_\delta$' or '$n \notin a_\delta$', which say, for each n, whether n belongs to a_δ or not. Each a_δ is then determined only by its members and the condition that these are elements of ω, not by any single, finitely expressible condition on its members.

If we were to add just one new set a, we could put

$$N_0 = L_0(a) = \omega \cup \{a\}$$
$$N_{a+1} = L_{a+1}(a) = Def(N_a) \cup N_a$$
$$N_a = L_a(a) = \bigcup_{\beta < a} N_\beta \text{ for } a \text{ a limit ordinal}$$

and put

$$N = \bigcup_{\beta < a_0} N_\beta$$

N will then consist of all the sets constructible from $\omega \cup \{a\}$. If N were a model of ZF, it would be one in which $V = L$ fails, since a is not constructible. But N, constructed in this way, is not guaranteed to be a model of ZF for any arbitrary a. The idea is that the sets added must be specified minimally, merely by an infinite list of the numbers which belong to them. This is because they must not contain any information about the minimal model M_{a_0}, such as the information that the ordinal number α_0 (the least upper bound of the ordinals in M_{a_0}) is denumerable, which can only be seen from 'outside' M_{a_0}. From 'outside' M_{a_0} we can see that α_0 corresponds to a well-ordering of ω and therefore to a subset of $\omega \times \omega$, and thence to a subset of ω itself. Neither this subset, nor the relevant correspondence between it and α_0 can belong either to M_{a_0} or to N, if N is to be a model of ZF. So some care has to be taken over how infinite subsets of ω are added.

The further idea is that no statements are to be true of any added a_δ which are not *required* to be true by its members or by the fact that it is a subset of ω. In other words, everything true of an a_δ in N must be 'forced' to be true on the basis of a finite amount of information about its membership together with the knowledge that any such finite amount of information can be extended. The basic tool is the definition of the 'forcing relation', defined as a relation between finite sets P of sentences, called 'forcing conditions', and sentences in the language of ZF. A *forcing condition* is a finite set of sentences of the forms '$n \in a_\delta$' and '$n \notin a_\delta$' which, for any given n and a_δ, does not contain sentences of both forms. The aim is to make precise the idea that a given finite amount of information about the new subsets of ω requires certain further statements in the language of ZF to be determined as true or false. So, for any forcing condition P

> P *forces* $n \in a_\delta(n \notin a_\delta)$ to be true if $n \in a_\delta(n \notin a_\delta)$ is one of the conditions in P.
>
> P *forces* $x \notin a_\delta$ to be true if x is not in ω.

Since no such conditions P have anything to say about the original members of M_{a_0}, it can be taken that for a, b in M_{a_0}, P forces $a \in b$ to be true if, and only if, $a \in b$ is true in M_{a_0}. In order to satisfy the axiom of extensionality, sets are required to be identical if, and only if, they have the same members. But each condition P gives only a finite and incomplete amount of information about the new infinite sets. This means that possible extensions of P also have to be considered. If two sets a_1 and a_2 really are non-identical, then there must be an element which is in one of them but not in the other. Thus it must be possible to find a sentence '$m \in a_1$' or '$m \in a_2$' to add to the list in P which will force the falsity of $a_1 = a_2$. So:

> P *forces* $a_1 = a_2$ to be true if, and only if, there is no extension Q of P for which Q forces $a_1 \neq a_2$.

The full definition of the forcing relation can be found in, for example, Cohen (1966) or Drake (1974). The inductive clauses of the definition closely parallel the inductive clauses in the definition of the standard satisfaction relation. Thus, for example,

> P *forces* A & B if, and only if, P forces A and P forces B.

But since the information in P is always incomplete, the trick is to find a way of getting a determination of the truth value of every sentence in the language of ZF extended by the addition of constants to name every element M_{a_0}, the new generic sets, and all sets constructible from these. Here the crucial clause is that for negation:

> P *forces* $\neg A$ if, and only if, for all Q extending P, Q does not force A.

Now although this is not enough to require that given any P and any A we must have P forces A or P forces $\neg A$ (since when P does not force A there might be a Q which extends P and does force A). It is clear from the negation clause that for every P and A there must be a Q extending P which forces either A or $\neg A$. (For the relations between forcing and intuitionist logic see Fitting (1969).) This means that in N, each statement about the new sets is decided by a finite amount of information about their membership. But it is not enough to have the truth value of every sentence settled by *some* Q extending P; it is necessary to have all sentences forced by Qs which are mutually consistent, or are all successive extensions of a single set of conditions. A *complete sequence of forcing conditions* is a sequence $\{P_n\}$ such that $P_{n+1} \supseteq P_n$ for all n and which is such that, for every sentence A in the extended language of ZF, there is an n such that P_n forces either A or $\neg A$. It is possible to prove that a complete sequence exists, because M_{a_0} is denumerable which means that the number of sentences in the extended language is also denumerable. So the sentences A_i of the extended ZF language can be listed, and P_{n+1} defined inductively as any condition Q extending P_n such that Q forces either A_{n+1} or $\neg A_{n+1}$.

If $\{P_n\}$ is a complete sequence, then for each n and a, some P_n forces either '$n \in a$' or '$n \notin a$'. A subset of ω given by such a complete sequence is called a *generic set*. If just one new set is added, it can be shown (a) that the sentences true in N and the sentences forced by the corresponding complete sequence of conditions of the forms '$n \in a$' and '$n \notin a$' coincide, and (b) that N does form a model of ZF. Furthermore, it is clear that V = L does not hold in N, but, by arguments closely paralleling those for L, it can be shown that AC and GCH do both hold. Thus AC and GCH are shown to be independent of V = L.

A model in which AC holds but GCH fails is obtained by adding a set of distinct generic sets a_δ, for $\delta < \mu$, where $\mu \geqslant \aleph_2$ in M_{a_0}. But, of course, from outside M_{a_0}, μ is denumerable. So the list of conditions '$n \in a_\delta$', '$n \notin a_\delta$', for $\delta < \mu$ will be denumerable. The idea is that the sets a_δ can all be forced to be distinct. No condition P can force $a_\alpha = a_\beta$ if $\alpha \neq \beta$, because there will be a condition Q extending P such that, for some n, either '$n \in a_\alpha$' and '$n \notin a_\beta$' or '$n \in a_\beta$' and '$n \notin a_\alpha$' are in Q. So in N the cardinality of $P(\omega)$ will have to be greater than $\aleph_1$. GCH will therefore be false in N. But since the a_δ form a well-ordered set, and the ordinals in μ are all in M_{α_0}, N will have to contain the pairs $\langle \delta, a_\delta \rangle$ and the set $\{\langle \delta, a_\delta \rangle: \delta < \aleph_\mu\}$. So in N the sets a_δ are well ordered. Since all other sets are constructible from these, the universe of N can be shown to be well ordered. AC can then be shown to hold in N.

In order to construct a model in which AC fails, infinitely many generic subsets of ω are added in such a way that any other set in the model can only distinguish between finitely many of them. Here the basic idea is that if AC were to hold, it would have to be possible to pick out a unique element from every set belonging to any given set of mutually disjoint sets and to form a set out of these selected elements. But infinitely many infinite subsets of ω will not all be distinguishable from one another on the basis of only a finite amount of information about the members of each. This has to be carefully exploited to produce an N in which the continuum is not well-ordered because it contains a subset T which, although infinite, contains no denumerable subset. T is the set of generic subsets a_m of ω given by a complete sequence of forcing conditions of the form '$n \in a_m$' or '$n \notin a_m$' for $m < \omega$. The a_m are all forced to be distinct because if m is distinct from k, then for any forcing condition P it will be possible to find a condition Q extending P such that Q forces '$n \in a_m$' and '$n \notin a_k$' (or vice versa) for some n. So T will be infinite. Suppose there were a condition P which forced it to be true that there is a function f from ω into T such that if $m \neq n$, $f(n) \neq f(m)$. P can contain information about only finitely many of the members of T. So there must be a $k < \omega$ such that whenever $\omega > s$, $t > k$, P contains no information about either a_s or a_t. But then there must be a P^+ extending P, such that, for some n, P^+ forces $f(n) = a_s$ and another extension P^* of P obtained from P^+ by interchanging s and t, leaving all other integers unchanged, such that P^* forces $f(n) = a_t$. But P^+ and P^* are compatible, since they are the same except that P^+ mentions a_s but not a_t, and P^* mentions a_t but not a_s. So $Q = P^+ \cup P^*$ is a forcing condition extending P. Q must then force f to be single valued, yet it also forces $f(n) = a_t$ and $f(n) = a_s$, which is impossible.

This proof of the independence of AC exploits the feature of it which renders it controversial – its assumption that a choice function exists independently of any means of making the choice, independently of any means of identifying the elements to be chosen. From the point of view of any finite, but indefinitely extensible, list of purely extensional information about infinitely many infinite sets there will always be sets which are distinct but currently indistinguishable. Hence there will always be sets between which it is impossible currently to force a choice. The plausibility of AC rests *either* on a purely extensional inter-pretation of the notion of 'set' (a set is just a collection of elements into a whole, without there being any implication about the existence, or the possibility, of a principle of selection), *or* on the fact that the notion of 'set' has already been so narrowly constrained that it can be demonstrated that there will be a principle available for making the required selections. It can, for example, be shown that GCH entails AC. This is because for GCH to hold there have to be plenty of one–one correspondences between sets. GCH requires that given any set A there cannot be a set B whose cardinality is in between that of A and $P(A)$. So, for any B such that A can be mapped one–one into B and B can be mapped one–one into $P(A)$, there must either be a one–one correspondence between A and B or between $P(A)$ and B. Intuitively speaking, what can be shown is that if all these mappings are available then there are enough to well-order every set. If AC fails, we have no reason to expect that the continuum can even be well-ordered, let alone that it can be assigned the initial ordinal $\aleph_1$.

9

Mathematical Structure – Construct and Reality

It is time to return to the questions raised in the Introduction. Should GCH be accepted as true? Are there exactly $\aleph_1$ points in a line? What basis is there even for supposing that these questions have, or can be given, any definitive answer? The framework which appeared to have given a precise sense to the question 'How many points are there in a line?', which determines the form that an answer should take, has been shown to be an insufficient basis for providing that answer. On what other resources can one draw? Or should it be concluded that this is simply a question which can have no definitive answer? The suggestions for getting an answer which have been canvassed are usually put into two groups (a) appeals to mathematical intuition, (b) appeals to mathematical consequences. However, as we shall see, at one level these are distinguished more by rhetorical emphasis than by the kind of mathematical grounds which would be adduced for justifying a decision on the continuum hypothesis.

1 Appeals to Mathematical Intuition

The requirement that the notion of 'set' should be restricted to 'constructible set', i.e. that $V = L$ be adopted as an axiom of set theory, is generally discounted as being both too restrictive and as having consequences which are in conflict with mathematical conceptions of 'set'. Thus appeal to $V = L$ to settle GCH is discounted on the basis of an intuitive and distinctively mathematical (as opposed to logical) conception of 'set'. Moreover, although Cantor believed CH to be self-evidently true, subsequent set theorists have not concurred. Thus in the conclusion to his book (1966), Cohen says:

A point of view which the author feels may eventually come to be accepted is that CH is *obviously* false. The main reason one accepts the Axiom of Infinity is probably that we feel it absurd to think that the process of adding only one set at a time can exhaust the entire universe. Similarly with the higher axioms of infinity. Now ω_1 is the set of countable ordinals and this is merely a special and the simplest way of generating a higher cardinal. The set C is, in contrast, generated by a totally new and more powerful principle, namely the Power Set Axiom. It is unreasonable to expect that any description of a larger cardinal which attempts to build up that cardinal from ideas deriving from the Replacement Axiom can ever reach C. Thus C is greater than $\aleph_n$, $\aleph_\omega$, $\aleph_\alpha$ where $\alpha = \aleph_\omega$ etc. This point of view regards C as an incredibly rich set given to us by one bold new axiom, which can never be approached by any piecemeal process of construction. (p. 151)

It is known that there are some restrictions on the possible values for the cardinality of C. König's theorem says that the cardinality of the continuum cannot be the sum of denumerably many smaller cardinals, and it can be shown that within ZF this is the only possible restriction on C. Generic models N can be constructed with

$C = \aleph_\alpha$ if α is not cofinal with ω in N
$C = \aleph_{\alpha+1}$ if α is cofinal with ω in N.

i.e. with $C = \aleph_2$, $\aleph_{\omega+1}$, $\aleph_{\omega 1}$, and so on.

In a similar vein, Drake (1974, p. 66) expresses the feeling that GCH is just too simple to be right. He gives the following supporting example: Given the cardinal number $\aleph_\alpha$, then $\aleph_{\alpha+1}$ is the collection of all isomorphism types of well orderings of $\aleph_\alpha$. It can be shown that $2^{\aleph_\alpha}$ is similar to the collection of all isomorphism types of all linear ordering of $\aleph_\alpha$. To say of a linear ordering that it is a well-ordering is a very strong requirement, so there should be many more linear orderings than well-orderings. Thus one should have

$2^{\aleph_a} > \aleph_{a+1}$. On the other hand he emphasizes the difficulty of proving the falsity of CH. At one time it was thought that the axioms asserting the existence of higher cardinals might provide a resolution of GCH. For example, the existence of a measurable cardinal entails $V \neq L$. But it was shown (by Levy and Solovay, 1967) that the existence of a measurable cardinal cannot decide the size of the continuum. It can also be shown that proofs of the consistency and independence of GCH can be modified to go through for ZF + axiom of inaccessibles. This is because CH is a problem which, if resolved at all, has to be resolved at a very low level of the cumulative type hierarchy. Since all that is at issue for CH is the existence or non-existence of a one–one correspondence between $P(\omega)$ (in $V_{\omega+1}$) and the set of all well-orderings of ω, its truth or falsity must be decided already at $V_{\omega+2}$ which is unaffected by the existence or non-existence of higher cardinals.

Here Drake is appealing to intuitions of the cumulative type hierarchy. It is within this structure that he constructs and discusses the various models of ZF. But what is the status of this hierarchy itself and how does one determine what questions are or are not resolved at a given level of it? Is there an intuitive grasp of this structure which outruns its characterization by the ZF axioms? At least one mathematician, namely Gödel, has been prepared to say that this is the case.

Given the proofs that AC and GCH are independent of ZF, one might well wonder whether there is any basis for determining the truth or falsity of CH. Might the situation in set theory not be analogous to that which obtains in geometry. There are, after all, certain historical parallels between CH and Euclid's fifth postulate – the parallel postulate. The postulate was initially adopted as self-evidently true. Then it was felt that it required proof, and finally it was shown to be independent of the remaining axioms of Euclidean geometry. The mathematical response to this was the creation of non-Euclidean geometries, in which the various ways in which the parallel postulate may fail are explored. It is no longer regarded as a sensible *mathematical* question to ask whether any of the axioms of Euclidean, or any other geometry, are *true*. If the question is asked as a question about the geometrical structure of physical space (or space–time) then it may be sensible, but it is a question for the physicist, not the mathematician and can only be answered by reference to physical theories and the body of empirical evidence to which such theories must answer. The mathematician may investigate the various geometries and characterize the structures in which their axioms are satisfied. Similarly CH was initially adopted by Cantor as obviously true; then he felt that proof was required. Finally it has been shown to be independent of the remaining axioms of set theory. Should that not mean that it should no longer be thought that there is just one set-theoretic universe to be uniquely characterized by the set theoretic axioms and by reference to which it can be sensibly asked whether CH or any other set theoretic claim is true or false? Should one not perhaps conclude that there are several set-theoretic structures, each of which can legitimately be explored by the mathematician?

In a well-known article, Gödel (1947) considers this suggestion only to reject it. He points out that in spite of the superficial similarities between the histories of CH and the parallel postulate, there are important differences between the situations in geometry and in set theory. In the case of set theory there are axioms, such as that asserting the existence of inaccessible cardinals, which although independent of the ZF axioms, are such that there is an asymmetry between systems in which it is asserted and ones in which it is denied. ZF + ¬(axiom of inaccessibles) has a model (e.g. the minimal model M) which can be defined and proved to be a model in ZF, whereas this cannot be done for ZF + axiom of inaccessibles. Moreover, in ZF + axiom of inaccessibles it is possible to prove the existence of a model for ZF. The addition of the axiom of inaccessible cardinals thus produces an extension of ZF in a much stronger sense than that in which the addition of the negation of this axiom would be an extension. In addition, in ZF + axiom of inaccessibles (but not in ZF + ¬(axiom of inaccessibles)) it is possible to prove new theorems about integers, whose individual instances can be verified by computation. CH can also be shown not to yield any new theorems in number theory and to be true in a model which is constructible within ZF. Gödel suggests that this might not be the case for some other assumption about the power of the continuum. However, Drake's remark that CH, unlike the question of inaccessible cardinal numbers, must be resolved at a very low level of the cumulative type hierarchy, suggests that its resolution one way or another will not produce the same kind of strong extension to ZF as is obtained by the addition of higher cardinal axioms.

Gödel also sees a contrast between the epistemological situations in geometry and set theory. In geometry the question of the truth or falsity of Euclid's fifth postulate retains its sense if the primitive terms are taken as referring to the behaviour of rigid bodies, light rays, etc. The question of the truth or falsity of CH similarly retains its sense if the primitive terms of set theory are read as referring to mathematical objects. And Gödel believes that it can retain a *mathematical* sense because he believes that sets are mathematical objects which have an independent existence.

> For someone who considers mathematical objects to exist independently of our construction and of our having an intuition of them individually, and who requires only that the general mathematical concepts must be sufficiently clear for us to be able to recognize their soundness and the truth of the axioms concerning them, there exists, I believe, a satisfactory foundation of Cantor's set theory in its whole original extent and meaning. (1964, p. 262)

The interpretation which gives them this meaning is outlined as follows:

> When theorems about all sets (or the existence of sets in general) are asserted, they can always be interpreted without any difficulty to mean that they hold for sets of integers as well as for sets of sets of integers, etc. (respectively that there either exist sets of integers, or sets of sets of integers, or . . . etc., which have the asserted property). This concept of set . . . according to which a set is something obtainable from the integers (or some other well-defined objects) by iterated application of the operation of 'set of', not something obtained by dividing the totality of all existing things into two categories, has never led to any antinomy whatsoever; that is, the perfectly 'naive' and uncritical working with this concept of set has so far proved perfectly self-consistent.

Since Gödel takes it to be this concept which has been axiomatized in systems such as ZF, he concludes:

> The set-theoretical concepts and theorems describe some well-determined reality, in which Cantor's conjecture must be either true or false. Hence its undecidability from the axioms being assumed today can only mean that these axioms do not contain a complete description of that reality. (p. 264)

So again, Gödel appeals to the basic structure of the cumulative type hierarchy as a hierarchy developed from the basis of an intuitive grasp of the integers. He further assumes that this hierarchy (a) has an existence which is independent of any axiomatic characterization of it, and (b) that it is determinate in all respects which can be expressed with the resources of the language of set theory, so that, in particular, CH must *be* true or false in this structure. Our problem, then, is an epistemological one – how to find out which is the case.

But by what means might we expect to be able to obtain, or to recognize a new axiom as giving us a more complete description of that reality? Intuitive self-evidence is one possible criterion, one which is fraught with the problem of explicating the relevant sense of 'intuition'. This is additionally problematic given that the rationale for the mathematical use of set theory was precisely to dispense with appeals to intuition. True this was carried out in the specific context where it was *geometrical* intuition which had proved inadequate as a basis for resolving problems in analysis. But if geometrical intuition, linked to the spatially depictable, is to be banished, in the name of rigour, it will not do simply to replace it by appeals to an altogether more mysterious set theoretic intuition which, in being divorced from the constraints of picturability, leaves the origin of its content wholly obscure.

However, Gödel also suggests that a decision about any proposed new axiom, or indeed the continuum hypothesis itself, might not be decided by its immediate and self-evident truth, but 'by inductively studying its success'. Here the source of the content of the supposed intuitions is indicated, but it is also the point at which the distinction between a decision on CH based on an appeal to mathematical intuition and a decision based on and appeal to mathematical consequence begins to blur.

2 Appeals to Mathematical Consequences

Gödel suggested that CH might be decided inductively by studying the success of either it or its negation. By 'success' he means something very close to what is meant in requiring success of a scientific theory – explanatory power or predictive success. It is a matter of having consequences which are already demonstrable by other means but where the new proof is simpler and easier to understand, making it possible to contract many different proofs into one proof. Here Gödel gives as an example the way in which the axioms for the system of real numbers are felt to be to some extent verified because they make it possible to prove number-theoretic results which can subsequently be verified in a more cumbersome way by elementary methods.

> A much higher degree of verification than that, however, is conceivable. There might exist axioms so abundant in their verifiable consequences shedding so much light upon a whole field, and yielding such powerful methods for solving problems (and even solving them constructively as far as that is possible) that, no matter whether or not they are intrinsically necessary, they would have to be accepted at least in the same sense as any well-established physical theory. (1964, p. 265)

The continuum hypothesis has consequences which Gödel finds highly implausible (such as there exist subsets of a straight line of the power of the continuum which are covered (up to denumerably many points) by every dense set of intervals ...). Yet there is no unanimity about what is and what is not plausible. Mathematicians are not all agreed in finding these results implausible. Here we see a problem about judging by consequences. If an analogy is to be drawn with the natural sciences, it should not be the plausibility or implausibility of predictions which determines a result, but the ability or inability to confirm these predictions experimentally. In mathematics there is no strictly experimental domain, but there are still domains which are distinct *vis-à-vis* their proof techniques.

Results about particular numbers reached by computation are results for which no further evidence can be required and which rest on no body of theory beyond elementary number theory. Indirect proofs of number theoretic results which go via additional set theoretic axioms can in some cases be 'confirmed' by the computational checking of instances. More generally, one could say that an appeal to mathematical consequences could only settle CH if either CH or its negation, or some more powerful set theoretic axiom which entails one or the other, could be shown to have consequences which can be either proved or disproved by more direct means. But if this is the case, it is far from clear what would be the correct philosophical conclusion to draw from such a determination. The price of drawing on an analogy with the empirical sciences is that of becoming embroiled in seemingly interminable debates between realist and anti-realist philosophies of science. The anti-realist would argue that the independent confirmation of the theoretical prediction is *not*, in this kind of situation, evidence of the *truth* of the theory used to make the prediction (inference to the best explanation is not a legitimate form of inference). Hilbert (1925), for example, expresses the views:

> The final test for every new mathematical theory is its success in answering pre-existent questions that the theory was not designed to answer. By their fruits ye shall know them – that applies also to theories.

and the view that every mathematical problem can be solved:

> One of the things that attracts us most when we apply ourselves to a mathematical problem is precisely that within us we always hear the call: here is the problem, search for the solution: you can find it by pure thought, for in mathematics there is no *ignorabimus*. (p. 384)

But this is in the context of a paper in which he is arguing that Cantor's paradise of transfinite numbers is to be preserved for the mathematician by showing the legitimacy of accepting these numbers as 'ideal elements', elements with no reality outside the formal structures developed for dealing with them. This was to be done by showing the reliability of their consequences within number theory. Hilbert's programme was to demonstrate the reliability of transfinite set theory by first formalizing the theory containing these ideal, infinite elements, and then providing a finitary, metatheoretic proof that no formula whose intended interpretation would be in conflict with number theory (such as '$1 \neq 1$') could be proved within the formal system. But from Gödel's incompleteness theorems we know not only that absolute consistency results for ZF or its extensions are unobtainable but also that if ZF is consistent then (a) it will contain formally undecidable sentences, and (b) there will be no consistency proof of it relative to elementary number theory.

In the absence of a consistency proof fulfilling Hilbert's requirements, and in the absence of any assurance that every mathematical problem has a solution, one can still ask what a 'test' of additional set theoretic axioms by its results in other domains might show. There are two senses in which the reliability of a system of ideal elements might be required: (a) That results proved indirectly can be proved, albeit in a more cumbersome way, directly (the ideal elements constitute a conservative extension of the theory of the original domain). In this case the 'ideal elements' are in principle eliminable and it would in principle be possible to introduce a plurality of different systems of 'ideal elements', each producing a different sort of simplification, or each simplifying different problems areas. (b) That the results proved indirectly cannot be proved directly but it can be shown in general that the system of 'ideal elements' is consistent and cannot lead to any contradiction with the theory of the original domain, so that it forms a consistent, but non-conservative extension of it. In this case, again, we have no a priori ground, other than an assumption of realism, for assuming that the theory of a domain can consistently be extended in only one way. Whereas the relation required under (a) would legitimate a fully anti-realist position *vis-á-vis* the proposed ideal elements, since their, in principle, eliminability would allow them to be regarded either as convenient fictions or as mere notational devices, that envisaged under (b) would result in a form of constructivism even for the original domain. For to allow for the non-conservative extension of the original domain in possibly more than one way (i.e. without basing this on realist assumptions about the antecedent existence of a unique extension) is to allow that what is true in this domain is not once and for all determined by its original form, but is determined in part by the way in which it is extended. The additional statements provable about the original domain cannot be regarded as having been antecedently true or false, but they become true or false in one or other extension; i.e. the original domain is something which either has the status of a construct to which more can be added, or contains indeterminacies which can be further determined.

So, whereas appeals to mathematical consequences may settle, or at least be an important factor in determining, the mathematical acceptability of a new theory, the philosophical conclusions to be drawn from the procedure of resolving an issue in this way are far from clear. They will depend crucially on the attitude taken toward the consequences themselves and to the mathematical domain which they concern. Are they taken as expressing truths about an independently existing domain or do they represent a new constructive determination of that domain? Are they results which can be established directly, if less elegantly, without going via the new theory?

3 Descriptive Set Thoery

The obvious place to look for consequences of set-theoretic axioms is to their impact on the point sets of analysis, for it was the task of classifying and describing these sets which first led Cantor to introduce his transfinite numbers. As Hallett (1979) shows, the acceptance of Cantor's transfinite set theory was indeed based on its consequences. He gives two examples of rather different kinds. One is Borel's 1895 proof of the Heine–Borel theorem (any countable cover of a closed interval by open intervals can be reduced to a finite cover). Borel's original proof makes use of countable transfinite ordinals, but his result was then soon proved using point set analysis alone and without making use of transfinite numbers. Borel himself had philosophical objections to transfinite numbers and to the actual infinite and was thus motivated to find a proof which did not employ them. Nevertheless he was prepared to acknowledge the heuristic importance of Cantor's methods. The other is a theorem about point sets which was proved in 1884 by

Mittag-Leffler. This made essential, non-eliminable use of Cantor's transfinite numbers. (Mittag-Leffler's theorem was 'Let Q be an isolated point set (i.e. $Q \cap Q^{(1)} = \varnothing$). Then it is possible to construct an analytic expression $F(z)$ such that the singularities of F are just the points of $Q \cup Q^{(1)}$.) This theorem enabled Mittag-Leffler to make considerable progress on the representation problem (see chapter 4, p. 79). This problem had been stated and recognized as important prior to Cantor's work and the fact that Cantor's theory could be applied in such a way as to make substantial progress toward a complete resolution of it could thus be interpreted as a vindication of the theory of transfinite numbers and one in which they could not be treated as a dispensable heuristic device. But they could still, none the less, be viewed as providing just an analytic tool facilitating the resolution of problems in point set theory rather than as forming any integral part of that theory. There is no immediately justified inference from utility to the adoption of any kind of realist attitude toward the theory found to be useful. However, subsequent work has shown that the connections between Cantorian set theory and the study of the point sets of analysis cannot be so easily disentangled, and indeed Cantor himself saw his work as an extension of that theory, not merely as an independent theory which might have application there.

The task of classifying and describing the point sets of analysis is carried out within what has come to be known as descriptive set theory. The descriptive set theorist is not primarily interested in the foundational project of reducing either the natural or real numbers to sets. He feels that his hold on the natural and real numbers is secure and in no need of set-theoretic underpinnings. That does not, however, mean that there will be no rôle for set theory. Its rôle is to assist the study of the structures of these domains. In particular, it is possible to define a hierarchy of sets of real numbers based either on the logical complexity of their defining propositional functions or on their construction from topologically simple sets.

We start with closed subsets of the plane (of $\mathscr{R} \times \mathscr{R}$). *Closed sets* are sets which contain all their limit points. For example, the set of points enclosed within a circle together with the points constituting its boundary would form a closed set. *Open sets* are the complements of closed sets (and contain a neighbourhood for every point which belongs to them). The class of *Borel sets* is the smallest class containing the open sets and closed under complementation and countable unions.

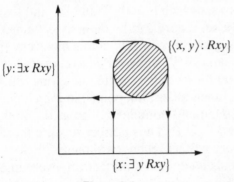

$\{y: \exists x\, Rxy\}$

$\{\langle x, y\rangle: Rxy\}$

$\{x: \exists y\, Rxy\}$

Figure 9.1

A Σ_1^1 or *analytic* set of reals is the projection onto the real line of a closed subset of the plane.

Alternatively, if atomic formulae have the form '$t_1 = t_2$' where t_1, t_2 are terms for either natural or real numbers, then a Σ_1^1 formula is any formula built up from atomic formulae and their negations using conjunction, disjunction, existential and universal quantification over the natural numbers and *existential* quantification over the real numbers.

A Σ_1^1 set is then a set which can be defined by a Σ_1^1 formula.

If the closed set, C, is defined as the set of points satisfying a given relation Rxy then the analytic sets obtainable from it by projection correspond to the two ways of defining sets of real numbers by introducing existential quantification over one or other of the free variables in Rxy. Thus existential quantification corresponds to the operation of projection (figure 9.1).

A Π_1^1 or *co-analytic* set is the complement of a Σ_1^1 set and will be defined by the negation of a Σ_1^1 formula. Since $\neg\exists x Ax \equiv \forall x \neg Ax$, projective sets will be definable by formulae which use only universal rather than existential quantification over the reals.

A Δ_1^1 set is a set which is both Σ_1^1 and Π_1^1, and which is such that both it and its complement have definitions in both quantifier forms. Such sets and their complements thus have the sort of logical homogeneity which was required by Russell's zig-zag theory.

A Σ_2^1 set is a set definable by a Σ_2^1 formula – one which is the result of applying existential quantification over the reals to a Π_1^1 formula.

A Π_2^1 is the complement of a Σ_2^1 set.

A Δ_2^1 set is a set which is both Σ_2^1 and Π_2^1, and so on.

Descriptive set theory can be defined as being the study of the mathematical properties of sets in this hierarchy (projective). For example Souslin (1917) showed that a set of reals is a Borel set iff it is Δ_1^1. A *perfect subset* of the continuum is one which is non-empty, closed (i.e. contains all its limit points, so P includes its first derived set $P^{(1)}$, $P \supseteq P^{(1)}$) and contains no isolated points, or is dense in itself, so that $P^{(1)} \supseteq P$. Thus perfect sets are those sets for which $P = P^{(1)}$. Every perfect set has cardinality $2^{\aleph_0}$. So if every uncountable set of reals contained a perfect set, every uncountable set of reals would have cardinality $2^{\aleph_0}$ and there would be no sets of real numbers with a cardinality in between $\aleph_0$ and $2^{\aleph_0}$. This would strongly suggest that $2^{\aleph_0}$ must be the next cardinal number after $\aleph_0$. Thus, Cantor had hoped to prove his continuum hypothesis by proving that every uncountable set of real numbers has a perfect subset. Lusin (1917) showed that every uncountable analytic set has cardinality $2^{\aleph_0}$. It can also be proved that every uncountable Borel set contains a perfect subset and thus has cardinality $2^{\aleph_0}$. However, it has also been shown using AC, that if the continuum can be well ordered then there is a decomposition of the real numbers into two, disjoint, uncountable sets A and B such that neither A nor B contains a perfect subset. This put paid to the idea of trying to prove the continuum hypothesis by trying to show that every uncountable set has a perfect subset. For if the continuum hypothesis were true it would mean that the continuum could be well ordered, and thus would contain uncountable sets containing no perfect subsets. In other words, there are questions about the real line which arise naturally from the context of analysis but whose answer can be shown to depend on the particular model of set theory being used.

Thus there is also in descriptive set theory a concern to determine what sets exist, or should be thought to exist, but the basis for determining this and for suggesting restrictions is not the same as for logicians. We have to remember that it is the attempt to represent a continuum as a continuum of points, something which seems to be both legitimated by and required by the algebraic notation for expressing functions of continuous variables, that makes use not only of the actual infinite, but of the non-denumerably infinite necessary. To get sufficiently many points to represent a continuum, we must, it seems, have $2^{\aleph_0}$ something of the cardinality of the power set of the natural numbers. But it is precisely the power set axiom which introduces an element of indeterminacy into the ZF axioms. In different models the set of all subsets of ω is different and it is this which is exploited in producing the independence proofs outlined in chapter 8. It is therefore not unnatural to look for ways of restricting the ZF universe so that the full power set axiom does not allow in sets of real numbers which are, from the point of view of analysis, pathological.

An axiom which has been suggested is the axiom of determinacy. Determinacy is a property of sets of real numbers which is defined as follows: Let x be a subset of the unit interval. Imagine a game for two players, I and II, in which players I and II take turns in selecting 0 or 1 until they have generated an infinite sequence $\langle s \rangle$. $\langle s \rangle$ can then be viewed as a real number in binary notation (or a subset of ω). If $\langle s \rangle \in X$ player I wins, otherwise player II wins. X is determined if either player I or player II has a winning strategy, i.e. a way of playing that results in a win for him no matter how the other player plays. A function f such that for every infinite sequence $\langle a \rangle$

$$\forall n(a(2n) = f(\langle a \rangle_{2n}) \Rightarrow \langle a \rangle \in X$$

would give a winning strategy for I. A function g such that for every infinite sequence $\langle a \rangle$,

$$\forall n(a(2n + 1) = g(\langle a \rangle_{2n + 1}) \Rightarrow \langle a \rangle \notin X$$

would give a winning strategy for II.

For example, every countable set of real numbers is determined, because given any countable set X of real numbers, all player II needs to do is to enumerate the set, and then for each of his turns put 0 if the $(2n + 1)$th place of r_n is 1 and 1 if it is 0. The result will then be a sequence which differs from each r_n in at least the $2_n + 1$th place, and which is therefore not in X.

The axiom of determinacy (AD) is the statement that every set of real numbers is determined. AD entails that every set of real numbers is either countable or has a perfect subset. Whereas AC entails the existence of sets of real numbers which are neither countable nor contain a perfect subset.

There have been two kinds of response to AD:

1 AD is an alternative to AC whose truth or falsity is unknowable, but whose consequences it might be interesting to investigate.
2 Since AD contradicts AC it is not plausible as an axiom of set theory. But there might be some inner model of ZF or its extensions in which some such axiom holds. The axiom of definable determinacy (DD) suggests that L[$\mathscr{R}$], the class of sets constructible from the reals satisfies AD.

The reasons given for thinking that DD should be accepted are of three kinds:

1 Determinate sets of reals are relatively well behaved – for example all uncountable determinate sets contain a perfect subset.
2 The axiom has consequences for the distribution of certain structural properties of sets in the projective hierarchy that are felt to be more plausible than those obtained from $V = L$.
3 DD yields direct and elegant game-theoretic proofs of some results which are provable in more complicated ways without it. (For further details see Maddy (1984).)

Be this as it may, DD will not yield a resolution of CH; since DD is designed not to conflict with AC it concerns only the class of sets constructible from the reals, not the totality of reals themselves, which it does not restrict in any way, nor the full power set of the reals. AD, on the other hand, conflicts with AC, and without AC we lose our justification for assuming that every set can be well-ordered and hence for assuming that there should be a determinate answer to CH. Moreover, there are other grounds for thinking that determining the consequences of additional set-theoretic axioms for the projective hierarchy will not give evidence which can confirm or refute CH.

It can be proved that:

Every Π_1^1 set is the union of $\aleph_1$ Borel sets.
Every Σ_2^1 set is the union of $\aleph_1$ Borel sets.
So that every uncountable Σ_2^1 set has the power $\aleph_1$ or the power $2^{\aleph_0}$.

But it is known that this result cannot be improved upon because $V = L$ entails the existence of an uncountable Π_1^1 set of reals with no perfect subsets, so that since $V = L$ also entails CH, CH is consistent with its not being the case that every uncountable projective subset of the reals has a perfect subset.

Further the axiom of measurable cardinals, MC, (which asserts the existence of a measurable cardinal) is a large cardinal axiom which does have consequences for the sets in the projective hierarchy.

MC entails that every Σ_2^1 set of reals has a perfect subset.
MC entails that every Σ_3^1 set is the union of $\aleph_2$ Borel sets and hence that every uncountable Σ_3^1 set has the power $\aleph_1$, $\aleph_2$ or $2^{\aleph_0}$.

In other words, the projective hierarchy on its own cannot provide evidence for CH since, in the presence of MC, it suggests the negation of CH. So, an appeal to descriptive set theory seems unlikely to yield a decision on CH. But perhaps more importantly, starting from descriptive set theory, it is not evident that there ought to be a determinate answer to CH. This is because with AD it is the status of AC which becomes more important. With AD, i.e. without AC, CH loses its significance – the question of determining the number of points in a line (the number of reals) is perhaps not the way to answer questions about the kinds of structures that can be imposed on the reals – the kinds of subsets of reals there can be. At this point we should take stock of the situation.

4 Foundations and Superstructures

One reason for insisting on a distinction between appeals to mathematical intuition and appeals to consequences as routes to reaching a decision on additional set-theoretic axioms is that they would not both be appropriate to the conception of set theory as providing a logical *foundation* for mathematics. To claim this status for set theory it is necessary to claim an independent and intrinsic justification for the assertion of set-theoretic axioms. It would be circular indeed to justify the logical foundations by appeal to their logical consequences, i.e. by appeal to the propositions for which they are going to provide the foundation. Thus, so long as set theory is cast in a foundational rôle, the appropriate basis for asserting an axiom is an appeal to set-theoretic intuition. It is, moreover, the conception of set theory as a foundation, an ultimate court of appeal in matters mathematical, that inspires the quest for addition axioms to give a more complete characterization of the set-theoretic universe. For it is part of the conception of a logical foundation that it be a precise, fixed, secure given starting point; there should be no unclarity over the notions taken as fundamental, especially that of 'set'; it is not the place where there is scope for creative activity. In other words, the conception of a logical foundation is inevitably realist and actualist. As was seen in chapter 7, even the claims of Frege and Russell to the effect that arithmetic, or the whole of mathematics, can be reduced to logic require both that a realist attitude be adopted toward logic and that the logic itself be appropriate to a realist and actualist view of the universes of both logical and non-logical objects. When it becomes clear that the mathematical notion 'set' cannot be reduced to the logical notion 'class' the claims for axiomatic set theory as a foundation have to be grounded in some basic kind of set-theoretic intuition.

This course would not be welcome to a logicist motivated by Russell's empiricist concerns. His concern was to show that mathematics is not grounded in some special non-empirical intuition, whether geometrical or arithmetical, of abstract, non-concrete objects; to show that it can have no claim to the status of being a body of synthetic a priori knowledge. His line of retreat from the reduction to logic as a *foundation* is to a form of conventionalism, where axiom systems are viewed as definitions of kinds of structures. On this view mathematics tells us merely about what logically follows from the axioms, but says nothing about whether there is anything of which these axioms are true. Once the reduction of 'set' to 'class' is abandoned, there is no reason to privilege set theory as foundational since other mathematical theories can be studied directly, providing that they are axiomatized. In this spirit various set theories can be explored, but there can be no question of expecting there to be an unequivocal determination of AC or GCH. Different extensions of ZF will constitute different definitions of 'set'. One might find some definitions more useful than others, but ultimately utility will be judged by reference to non-mathematical applications, not by application strictly within mathematics.

The standpoint of descriptive set theory, by contrast, departs from the view of set theory as foundational in a different way, for it rests on the view, also expressed by Gödel in the quotations on p. 196, that the mathematicians' hold on the natural and real numbers is secure and in no need of set-theoretic underpinnings. From this point of view set theory is to be judged by reference to its consequences for these mathematical domains. But if this is so, how are we to view the mathematical rôle of set theory? What attitude is to be taken to CH, given that it seems unlikely that descriptive set theory will provide any direct basis for decision? One might suggest an inversion; set theory, far from being a foundation, is a superstructure and the rôle of a superstructure is to provide a 'space' within which to construct structures, a space which is to be the bearer of all possible mathematical structures. It would then be essential to a superstructure that it *not* set determinate bounds on the notion of a possible structure, that this should be open to further determination in the mathematical future. This might be the lesson to be drawn from this story of the way in which classical finitism was displaced and in which CH came to be proposed, was given sense, and then turned out to be neither provable nor refutable, and also to be unlikely to be decided either by appeal to set-theoretic intuition or by appeal to consequences? So let us briefly recapitulate on that story.

First we have seen (in chapter 1) that prior to the algebraization of geometry the classical finitist could stand his ground without suffering mathematical penalties. So long as geometry (the science of continuous magnitudes) and arithmetic (the science of discrete magnitudes) remained separate and largely independent branches of mathematics, the infinite inherent in the continuum remained internal to the notion of continuity which was taken as a primitive notion grounded in geometric intuition. In this context the question

'How many points are there in a line?' has no mathematical sense, even if the actual infinite is allowed, because a line as a continuous magnitude does not form the sort of whole to which a number, other than a measure of length, is assignable. A continuum cannot be made out of points (is not a set of points) even if it is granted that it can be actually infinitely divided into infinitesimals. A continuous whole is a whole given 'before' its parts, parts which are the product of division, where points are the boundaries introducing divisions into a line. A precondition of being able to answer a 'How many?' question is that the question be asked about a determinate totality of discrete entities. Thus a first condition of giving sense to the question about the number of points in a line is that it be possible to regard a line as a determinate totality of points, as a whole given 'after' its parts.

However, as argued in chapter 2, the classical finitist can stand his ground without mathematical penalty only on condition that he does not embrace nominalism or its accompanying extensional view of classes. For the finitist mathematician needs to be able to talk and reason about all members of a potentially infinite class, such as the class of all (possible) triangles, or of all (possible) points of division of a line, without commitment to the existence of these possibilities as members of a determinate totality, which would be an actually infinite class. So long as the logical framework which informs the view of classes remained centre round the Aristotelian theory of the syllogism, which was ambiguous between extensional and intensional readings, discrepancies over the status of the infinite as actual or potential do not come to the fore; the notion of class could remain ambiguous.

The situation changes, however, with the algebraization of geometry and the development of the important notion of a mathematical function. As outlined in chapter 4, the concept of a function develops away from its geometrical origin, in the algebraic expression of the 'law of motion' generating a continuous curve, as the algebraic techniques come to take over from purely geometrical methods. With the introduction of an algebraic notation there is immediately a tension between the view of a continuous curve as a whole given before its parts and as a whole composed of discrete entities – given after its parts – for the use of variables in expressions such as $y = ax_2 + bx + c$ suggests that 'x' and 'y' are to be read like '*a*', '*b*' and '*c*', as standing for numbers. However, the very term 'variable' indicates an alternative understanding (illustrated in the quotation from Newton on p. 76) on which a variable is read as standing for a changing magnitude so that the equation expresses the way in which magnitudes co-vary. A graph is itself just a way of illustrating what a function expresses – a structured relation between two or more independently specified magnitudes (which need not be spatial magnitudes). The mathematician's concern is with the structure of this relationship, not with the specific character of the magnitudes related; with obtaining a mathematically perspicuous representation of a complex relational structure, not with its possible concrete instantiations. So long as variables could stand ambiguously for continuously varying magnitudes or for individual magnitudes, or for numbers, and so long as geometric and algebraic representations were used in tandem it remained possible to avoid the issue of whether a continuum has to be thought to be made up of points. It is, however, with the shift of the focus of mathematical concern away from geometric figures and toward relational structures, which begins with the algebraic characterization of the traditional object of geometry, that mathematical reasoning decisively breaks the bounds of any rational framework relying only on the categories of Aristotelian syllogistic logic, which is incapable of handling relations effectively.

In chapter 4 it was suggested that, from the point of view of the tenability of classical finitism, the crucial step comes with the acceptance of the 'pathological' functions as legitimate functions, legitimate objects of mathematical study. From a purely abstract point of view it would have been possible to refuse to count such things as functions (cf. the discussion of monster-barring in Lakatos (1963)). This was not, however, a practical possibility given that functions expressed in very similar algebraic forms, i.e. involving possibly infinite expansions using the trigonometric sine and cosine functions, are vitally important to physics in the development of Fourier analysis and the mathematical methods for dealing with wave phenomena. The pathological functions demonstrate decisively that the powers of algebraic representation outstrip those of an intuition based only on the geometric continuum. The classical finitist could accept the graphs of continuous functions in the same way that he accepts unbounded straight lines, i.e. as potentially infinite, always actually finite but indefinitely extensible. These graphs do not need to be considered as sets of points. But the graph of a function which has infinitely many discontinuities in any given arbitrarily small interval (see p. 81) is not only not picturable, it also immediately introduces an actual infinity of points of division. It is this actual infinity of points of division which renders it impossible to draw or to imagine the graph.

Here, it seems, the classical finitist is faced with a choice. Either (a) he must throw away geometrical intuitions and the geometrical continuum, resolving to do all his mathematics on an arithmetic, algebraic basis, where the only source of a notion of the potential infinite is that associated with the indefinite extensibility of the natural number series. Or (b), if he retains geometrical intuition and the geometrical continuum, he must admit that the introduction of algebraic methods into geometry commits one to the conception of the continuum as actually infinitely divisible and that these methods also provide the vehicle through which sense is to be made of this conception. Option (a) involves an impoverishment of the foundations of his mathematical world, and will lead him to follow the course of intuitionist mathematicians, constructing non-classical analysis. Whereas geometric intuition would have convinced the classical finitist that the limit of a convergent infinite sequence of rational numbers exists independently of the sequence approximating it, if cut off from that intuition he will have to say that an infinite sequence has a limit means only that it converges; there is no independently existing limit to which the sequence is a sequence of approximations and there can be no completed, actually infinite sequence. Thus all statements about a potentially infinite sequence must, and can only, be made on the basis either of knowledge of the law which generates them or on the basis of a finite segment of the sequence. This does not justify assertion of the law of excluded middle in relation to statements concerning infinite sequences or concerning the limits of those which converge. In other words, real numbers will not in general be fully determinate objects and it becomes necessary to develop an alternative to classical analysis. From the intuitionist point of view, classical analysis cannot be regarded as being logically well founded.

Taking option (b) and retaining reliance on geometric intuition, however, involves abandoning his finitism once he is shown that it is possible to think coherently about the line as an actually infinite totality of points and is shown the mathematical reasons why such a conception should be adopted. Now a classical finitist who chose, on these grounds, to abandon his finitism would be doing so on the ground that it is the infinite inherent in the geometric continuum which itself leads him to the view. That is, he would have taken the second of the above options because he was reluctant to abandon geometrical intuition and the kind of mathematics traditionally based on it. It is the mathematical exploration of this continuum and of the kinds of structures it can be conceived (not pictured or imagined) to have imposed on it that will have persuaded him that he will have to think in terms of actually infinite division, and which will have persuaded him of the need to construct an intellectual model of the linear continuum which will make sense of the notion and will provide the tools for further mathematical characterization of the possible structures realizable within a continuous space.

Here he would be in accord with Dedekind's remarks (quoted on p. 85) about the construction of the real numbers; they are an instrument to be constructed for the purpose of arithmetically modelling the line, so providing an analytic tool for explicating this intuitively simple notion of continuity. Dedekind's definition of real numbers as 'cuts' in the set of rational numbers is thus motivated by geometric intuition. Although his cuts are defined set theoretically (a cut is an ordered pair $\langle A, B \rangle$ of sets of rational numbers which is such that $A \cup B$ is the entire set of rational numbers and every number in A is less than every number in B) they are seen as representations of the points of division of a line. Dedekind introduced irrational numbers to correspond to those cuts not produced by any rational number (for example when A consists of all rational numbers whose squares are less than 2 and B consists of all the remaining rational numbers the cut is produced by $\sqrt{2}$) but they are treated as being wholly defined by the cuts to which they correspond. Russell castigated Dedekind's procedure of postulating irrational numbers as productive of cuts:

> The method of 'postulating' what we want has many advantages; they are the same as the advantages of theft over honest toil. (Russell, 1919, p. 71)

But the concerns of Russell and Dedekind are different. Dedekind is trying to model the geometric continuum arithmetically, where Russell is seeking to eliminate it entirely and to show that mathematics (including geometry) can be reduced to logic.

If our erstwhile classical finitist has been persuaded that the real numbers do provide an arithmetic model of the linear continuum, one in which it is modelled as a set of points, he might also be persuaded to accept Cantor's transfinite numbers if they prove to be effective instruments for the further investigation of the complexity of the continuum. Let us briefly review the remaining steps by which sense came to be given to the question about the number of points in a line before asking how compelling they would be to the classical finitist who does not want to follow the course of intuitionism.

With the advent of the real line, the arithmetic point continuum modelling the geometric line, the question of how many points there are in a line is translated into the question 'How many real numbers are there?' and this looks as if it might make some sense, since the real numbers do at least form a totality of discrete entities. None the less there remains the problem that this is an infinite totality and is thus 'too large' to be counted. Furthermore, such 'large' totalities can be put in one–one correspondence with proper parts of themselves. This can be looked at in one of two ways: *either* these totalities are such that different ways of counting produce different answers (different ordinal numbers) so that there is no one answer to a 'How many?' question, *or* they are such that there is a unique answer but a part may have just as many members as the whole. In the finite case (correct) counting always, necessarily, produces the same results and whole sets always, necessarily, have more members than their proper parts. If infinite totalities are to be numbered, if there are to be transfinite numbers, it has to be allowed either that these principles are not essential to the concept of number (are not really necessary truths) or that our concept of number should be modified and generalized to cover situations in which these principles do not hold.

Until Cantor proved that there could be no one–one correspondence between the real numbers and the natural numbers it was not clear that there was any way for the concept of number to get a grip in the transfinite – there could only be the single classification 'infinite'. But now one of the criteria already in use for asserting numerical difference is seen to yield the conclusion that there are not as many natural numbers as there are real numbers; infinites are not, apparently, all of the same kind and there is some basis for thinking that the difference has a claim to be regarded as a difference in numerical magnitude. This gives rise to the separation of the notion of cardinality (linked to the existence of one–one correspondences) from that of ordinality (linked to a generalization into the infinite of the counting process and to the existence of well-orderings). Each infinite set may have different ordinal numbers (depending on how it is counted (well-ordered)) but only one cardinal number, which is defined to be the least of its possible ordinal numbers. The introduction of transfinite ordinal and cardinal numbers thus gives further sense to the question about points in a line; it becomes the question 'What is the least transfinite ordinal number which can be given to the set of real numbers?' or, since it can be shown that there is a one–one correspondence between the set of real numbers and the set of subsets of the natural numbers, 'What is the least transfinite ordinal number which can be given to the set of all subsets of the natural numbers?' This then becomes a more precise, but still seemingly unanswerable, question, when the theory of transfinite numbers is grounded in axiomatic set theory. At this point does the question have a fully determinate sense? That is, is it such that it should be thought to have a fully determinate answer even though the axioms so far adopted do not suffice to provide it?

What I hope to have made clear is that the question certainly does not have a sense in isolation. It has a sense only against the background of a framework within which the form of an answer (a set of possible answers) is defined. The erection of this framework is certainly not a matter of entirely free creative activity. Each addition to the sense previously attached to our question is constrained by the partial sense it already possesses. To say that it has *some* sense, is to say that one cannot invent any meaning one likes for it. To think that it already has a determinate content (and hence a determinate answer) which is being progressively revealed would be to think that the constraint amounts to determination – at each stage there should be just one objectively correct way of proceeding. Yet if one looks at these stages, this claim would be hard to justify.

In particular, the lack of one–one correspondence between the natural numbers and the real numbers, or between any set and its power set, is not inevitably to be interpreted as a *numerical* result. This interpretation is natural only after the introduction of the transfinite cardinal numbers. Further, even with the separation of cardinality from ordinality, cardinality is not justifiably claimed as a *numerical* property without appeal to the axiom of choice. But AC, like CH, is independent of the remaining ZF axioms and its status remains controversial. In the context of AC it is natural to assume that CH must *be* either true or false, even if we cannot know which is the case. This is because AC entails that the continuum can be well-ordered and hence that there must be a one–one correspondence between it and some ordinal number. But then there must be a least such ordinal number and this initial ordinal number will be the cardinal number of the continuum.

This illustrates the 'ontologizing' potential of AC; it can appear to represent a highly realist, extensionalist and non-constructive determination of the notion of a 'set'. But, as Cohen remarks in the passage quoted on p. 193, in ZF the power set of an infinite set is impredicatively defined and its membership is thus under-determined relative to the remaining axioms. The inclusion of the power set axiom thus leaves an indeterminacy in what is, from the point of view of ZF, to count as a set. It is this under-determination which makes it possible to produce models of ZF in which CH holds and models in which it fails, and also renders the force of AC indeterminate. If 'set' were restricted to 'constructible set' (i.e. $V = L$), then clearly AC holds, and within this context it does not require grounding in any strong realist or extensionalist assumptions. It is only when *any* arbitrary collection of natural numbers can constitute a subset of ω, even though there may be no means of specifying it, that AC comes in a parallel way to express a highly realist view, asserting the existence of choice sets in the absence of any ground for supposing that there is any principle in which the choice might be based. From a philosophic point of view, then, if not from the point of view of strict entailment in ZF, there is a close connection between the status of AC and that of CH. For example, it might be suggested, as for AD, that AC holds within certain inner models of ZF (when the notion of 'set' is limited in some way) justifying this by the thought that the use of AC amounts to a tacit assumption that the notion of 'set' has been limited. When AC is used it is frequently possible to specify the kind of sets for which the assumption has been made.

The presumptions grounding the supposition that CH has, and ought to have, a determinate answer are thus (a) that the subsets of ω form a determinate totality and (b) that this totality can be well-ordered. It might therefore be argued that anyone wishing positively to refuse the numerical question about the continuum, insisting that it lacks a determinate sense and therefore a determinate answer because it rests on false or unwarranted presuppositions, would be just as committed to thinking that the questions 'Do the points in a line form a determinate totality?' and 'Can the points in a line be well-ordered?' have determinate answers, as one who is committed to the belief that CH itself has a determinate answer. But within what framework and from what perspective might one suppose these questions to have determinate but negative answers?

From the point of view of the classical finitist they can both be answered negatively. No infinite totality is determinate, has a determinate membership. Moreover, in the case of the points in a line, whether these are conceived geometrically or arithmetically, via non-terminating decimals, one could not expect to well-order the points in a line because these points are themselves not precisely individuated from one another. We have no means of uniquely identifying every point in a line, or every real number, and hence have no means of ordering them in such a way as to be sure that every subset has a least element under this ordering. Either we construct a well-ordering and it is incomplete, or we define a relation over all possible real numbers and find that it is not a well-ordering. This is merely a reflection of what Zeno's arguments show – that a continuum cannot be built up out of points (not even a point continuum). The intuitionist, who remains committed to the view that the only possible notion of the infinite is the potential infinite, would endorse this view.

But what of our classical finitist who has become convinced, on the basis of geometrical intuition, to abandon his old style classical finitism and accepts the standard, as opposed to the intuitionistic, theory of real numbers? He accepts this because he accepts the need for an 'arithmetical' model of the continuum to supplement, but not to displace, geometrical intuition. This 'arithmetical' model will have to be conceived as containing an actual infinity of points. The fundamental tenet of classical finitism that has been surrendered is the view that a continuum cannot be conceived as a set of points. However, this surrender is only made possible on condition that the point continuum is not conceived as being *constructed* or *made up out of* points, but is modelled as a set of objects (points) – the set of limits of convergent sequences of rational, or of cuts in the set of rational, numbers. It becomes possible to think of the continuum as a totality of points only in the context of a clear distinction between parts and members; points are not parts of the continuum they are members of it (cf. p. 15). This in itself need not require him to abandon the view that there remains a fundamental distinction between discrete and continuous wholes – something which should be reflected even when the continuum is supplied with a discrete model. For, from his point of view, the possibility of the discrete model is itself grounded in the geometrical continuum, it is not something which can be independently constructed, for without geometric intuition he would not accept that limits of infinite sequences of rationals exist as determinate objects. It is as a theory of continuous magnitudes (see Bostock, 1975, ch. 3) that he accepts the real numbers, not as set-theoretic constructs. A continuous whole remains for him a whole given before its parts (i.e. which cannot be made up from them) and a discrete whole is one which is given after its parts. This requires that the 'discrete' model of a continuum be regarded as a class whose membership is indeterminate – the class of real numbers is a class of discrete entities, but is only given as a whole intensionally, it cannot be conceived as extensionally determined when its extensional determination corresponds to a discrimination of the parts of the continuum.

Further, if we reflect on the rôle that the assumption of continuity plays in geometry, I think there is good reason to suggest that there ought to be an element of indeterminacy in any 'discrete' model of the continuum. The continuum of old-fashioned geometry was infinitely homogeneous and without structure; it is the mathematical counterpart of prime matter – that which is capable of receiving any and all forms but which is itself formless. This is what infinite divisibility amounts to – the continuum can be cut *anywhere*, a boundary line can be drawn anywhere in space and can assume any shape. The continuous spaces of twentieth-century geometry and topology can no longer be regarded as structureless – even continuous spaces have their structural properties; they can be Euclidean or non-Euclidean, orientable or non-orientable, have an intrinsic curvature, be non-homogeneous in certain topological respects, etc. Continuity itself is not a simple unanalyzed, given concept, but one which has been subjected to analysis in Hilbert's axiomatization of geometry. It is possible to conceive of 'continua' which are homogeneous in the sense of being 'structured all the way down'. For example consider the following line (derived from the Cantor discontinuum). We start with the closed unit interval $[0, 1]$ and divide it equally into thirds erecting an equilateral triangle on the centre third and rubbing out the base (removing the points in the open interval $(\frac{1}{3}, \frac{2}{3})$ (figure 9.2(a)). We then repeat the process, erecting equilateral triangles on the centre thirds of the two intervals $[0, \frac{1}{3}]$ and $[\frac{2}{3}, 1]$ and rubbing out the open intervals $(\frac{1}{9}, \frac{2}{9})$ and $(\frac{7}{9}, \frac{8}{9})$ (figure 9.2(b)). We now allow the process to be repeated indefinitely. So that there will be no flat bits left on the graph no matter what level of magnification, going down in powers of 3, we reach, we will find that there is an equilateral triangle forming a hump over the central third of the interval. This zig-zag line intersects the original straight line in non-denumerably many points and these form what is known as the Cantor discontinuum, because the points are such that there is always a 'gap' between any two of them. Even so it remains part of the notion of a continuous space that it is that in which all possible structures, all possible patterns of discontinuity may be introduced and studied. In this respect

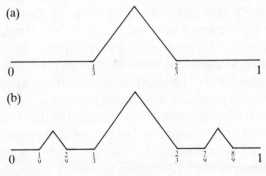

Figure 9.2

geometrical space is absolutely infinite (unlimited – cf. p. 26). We can and should have no way of giving any intensionally or extensionally precise characterization of all possible structures – to have this would be to foreclose on future mathematics.

From this standpoint, then, there would be a ground for refusing to accept that the question of 'How many points are there in a line?' makes sense, and hence for rejecting the idea that CH should have a determinate answer. But would it also mean rejecting the whole framework of transfinite set theory? I do not think so. The position outlined would be one which would be committed to taking seriously the claims of the theory of transfinite ordinal numbers, at least those in the second number class, in so far as they do give a means of characterizing the unpicturably complex structures which may be algebraically imposed on a continuum. Axiomatic set theory, in its own right, has also shown its value as a general framework within which structures may be mathematically characterized and studied.

The study of structure requires as a backdrop a superstructure, or a space – a universe projected as the abstract recipient of all possible structure within which structures of specified kinds are realized and studied. This superstructure will have to remain beyond complete mathematical determination – it cannot be an actual infinite in Cantor's sense. From this point of view one could see that there is a certain incompatibility between the conception of the universe of set theory as providing such a structure and the idea of tying it as closely as possible to procedures of construction which determine its structure and the structure of sets within it. The universe or space must be a whole given before it parts (and this is recognized in ZF with the distinction between classes and sets — the universe itself cannot be a set, but is a class, similarly for the ordinal numbers). The parts, like the points in a line, are in one sense presumed to exist, but in another sense do not exist prior to their discrimination or characterization – their existence is as potential or possible, rather than actual, objects.

It may then not be reasonable to suppose that the continuum can be well ordered or be 'numbered' because this would require it to have the sort of determinacy of structure which it is precisely the function of the continuum, whether geometric or as the set of all possible subsets of ω, to lack. This is to see a certain symmetry between the indeterminacy introduced into the mathematical notion of set via the power set axiom and the indefinite receptivity of structure of the geometrical continuum. There is no in principle limit that we can impose on our capacity to introduce and define new kinds of structure. The continuum cannot be numbered, not because the actual infinite makes no sense, but because the infinite associated with its continuity remains irreducible to fully extensionally determinate, discrete determination and it is important that it does so, but our conception of the continuum may change and be elaborated as we find new means of superimposing structures on it.

But if the continuum is to be without number, then the axiom of choice cannot be unrestrictedly asserted, yet mathematicians would be very reluctant to give up this axiom. The ground for thinking that it should be asserted comes in part from the history of the development of the notion of a function where it has come to be accepted that this should be liberalized to any many–one pairing of objects without the requirement that there be law or principle which effects this pairing. If this is allowed then why not a choice function for any set of mutually disjoint sets? Hasn't mathematics already shown itself to be inherently realist in its commitments?

The fundamental issue is then how the mathematician reasons about all possibilities prior to their specification. To give a theoretical discussion of real-valued functions requires not that one just consider specific examples, but also all possible real-valued functions. If no reasonable limit can be set on the means which might be available for defining these, then the notion of a possible real-valued function cannot be tied to, or restricted by any particular conception of the means for defining them, even though, at any given historical period this will *de facto* be the case. The same could be thought to go for sets or classes. The mathematical theory, if it is to be universally applicable, cannot be tied to any specific means for defining or determining these. However, this justification of mathematical procedure, whilst it suggests that mathematical theory will have to treat classes and functions as if they were given extensionally, independently of their defining conditions, also suggests that its own theoretical concepts cannot be subjected to the same treatment. The theory of all possible classifications over some domain, treated as given, must itself exploit the non-extensional route of delimiting the domain of the possible intensionally.

This is what makes the realism–antirealism issue difficult to resolve head on.

> It seems that we ought to interpose between the Platonist and the constructivist picture an intermediate picture, say of objects springing into being in response to our probing. We do not *make* the objects but must accept them as we find them (this corresponds to the proof imposing itself on us); but they were not already there for our statements to be true or false or before we carried out the investigations which brought them into being. (This is of course intended only as a picture; but its point is to break what seems to be the false dichotomy between the Platonist and the constructivist pictures which surreptitiously dominates our thinking about philosophy of mathematics.) (Dummett, 1959, p. 509)

The mathematician needs a framework, a superstructure, a representation of space, within which to work and this can only be postulated as a range of possibilities that is left indefinite. But it is as if these possibilities were there to be discovered. What is discovered is that something is possible and this is done by inventing a way of constructing or doing it; a determination is introduced which non-conservatively extends the theory. But there is also a hierarchy of frameworks where we can study from one the nature of the possibilities available within another by presuming that the framework studied does not exhaust the realm of all possibilities. The problematic status of set theory derives from the fact that it has tried to assume the double role of framework and of mathematical object – it is within set theory that the models of set theory are investigated. But the set theoretic universe cannot simultaneously be both object of study and framework within which to study.

Cantor was exploring the infinite possibilities inherent in, and constrained by, the indeterminacy of the 'formless' geometrical continuum but by methods which had the power to realize, or actualize, possibilities which could not have been realized previously. By assuming that he was revealing realities, actualities, rather than exploring possibilities (by setting up, creating, a framework for their representation) he was led to assume that the continuum has a determinate structure which could be characterized once and for all in such a way that one could see how it could be constructed (by an actually infinite being) out of points. But this assumes that one can obliterate the distinction between wholes given before and wholes given after their parts, between the actual and the possible. The subsequent history of set theory suggests that mathematics remains the art of the possible, of the potential, even when it has come to terms with the actual infinite.

Further Reading

Chapter 1   The Finite Universe

Rucker (1982) would provide an alternative, very informal but stimulating way into the problems of the infinite, written from a strongly realist perspective. Further examples of the paradoxes of the infinite discussed after Aristotle but which served to reinforce classical finitism can be found in Bolzano (1851). On Zeno's paradoxes and on the conceptual problems relating to space and time Salmon (1967) and Gale (1967) are useful collections. More sustained treatments can be found in Fraassen (1970) and Lucas (1973). The pressure toward treating space as an individual whole is admirably treated in the early chapters of Nerlich (1976).

Chapter 2   Classes and Aristotelian Logic

Histories of logic are few and far between. The standard reference is Kneale and Kneale (1962). Geach (1962) covers in detail some of the issues discussed in this chapter. Tamas (1986) gives a very thorough treatment of logical and related metaphysical issues as they appear in the basically Aristotelian framework. Putnam (1967) gives a good exposition of Russell's logicist views about mathematics.

Chapter 3   Permutations, Combinations and Infinite Cardinalities

Hacking (1975) provides an excellent historical introduction to the topic of probability and contains an extensive bibliography covering the period 1654–1700. Kneale (1949) remains one of the clearest introductory discussions of the various interpretations of probability. Hacking (1965) gives a more detailed discussion of theories of chance and their bearing on statistical inference, whereas Mellor (1971) contains a more philosophical discussion of how the notion of 'chance' should be interpreted.

Chapter 4 Numbering the Continuum

The story of this development is concisely and clearly related in Mannheim (1964). A less mathematically technical and very readable account of the development of the concept of 'number' to include real numbers is Dantzig (1954). Boyer (1959) provides a more general history of calculus and is a useful starting point although more recent historical work would query, or at least place caveats on, his interpretations of the period from the Greeks to the seventeenth century. Good examples of such recent discussions are North (1983) and McGuire (1983). Carruccio (1964) covers the same ground as Boyer but from a rather more philosophical perspective. Vilkenkin (1968) contains some marvellous examples of the curious functions and sets that emerged as the actual infinite came to be taken seriously in mathematics.

Chapter 5 Cantor's Transfinite Paradise

There are two relatively recent books which between them provide an excellent coverage of Cantor's work. Dauben (1979) is a mathematical and philosophical biography of Cantor, and Hallett (1984) is a very thorough discussion of Cantorian set theory set against the background of more recent work. In order to have an idea of a contrasting approach which might have been adopted it is worth looking at Dedekind (1901).

Chapter 6 Axiomatic Set Theory

A good selection of papers written at the time of the axiomatization of set theory, including papers written by Zermelo, Skolem and Fraenkel, can be found in Heijenhoort (1967). There are very many good mathematical texts on axiomatic set theory. An introductory text which goes directly to the theory of transfinite numbers is Rotman and Kneebone (1966). A very much more complete and more sophisticated text would be Jech (1978). A brief run down on the various axiomatizations can be found in Hatcher (1968, chs 5 and 7), and also at the end of Quine (1967).

Chapter 7 Logical Objects and Logical Types

Frege (1959) is itself a very readable introduction to Frege's views on arithmetic. Dummett (1967) gives an excellent brief account of Frege's philosophy which pays particular attention to his philosophy of mathematics. Wright (1983) gives a more detailed treatment. An excellent exposition of one of the positions to which Frege was opposed is given in Tragesser (1984). Russell's various proposed approaches to the resolution of set theoretic paradoxes are set out in Russell (1907). Russell's notion of zig-zag sets is imaginatively discussed under the heading of 'figure and ground' in Hofstadter (1979, ch. III). The clearest exposition of the ramified theory of types is in Russell (1908) and the simple theory can be found in Russell (1919). Two critical discussions of Russell's position, both written from a standpoint of realism about sets, are Ramsey (1926) and Gödel (1944). Chihara (1973) gives a very helpful discussion from a standpoint more sympathetic to Russell. Frege's system and theories of types are set out in Hatcher (1968, chs. 3 and 4). The iterative conception of set is discussed in Boolos (1971) and Parsons (1977).

Chapter 8   Independence Results

A very complete technical treatment of set theory and the independence results can be found in Jech (1978). Drake (1974) covers independence results with special reference to consideration of the addition of large cardinal axioms; throughout, the discussion is set against the background of the cumulative type hierarchy. Cohen (1964) remains worth reading for his perspective on set theory, even though the results have since been more elegantly presented. A more comprehensive and elegant treatment of independence proofs can be found in Bell (1977), where both forcing and the approach through Boolean valued models are covered. Fitting (1969) is interesting for its exploration of the connections between forcing and intuitionistic logic.

Chapter 9   Mathematical Structure – Construct and Reality

The second half of Hallett (1984) contains an evaluation of the import of independence results for the original Cantorian conception of set theory. Maddy (1984) contains a useful discussion of the issues raised by the appeal to descriptive set theory and the axiom of determinacy. A more detailed account of descriptive set theory can be found in Mansfield and Weitkamp (1985).

Bibliography

Aristotle (1984) J. Barnes (ed.) *The Complete Works of Aristotle*, Princeton University Press/Bollingen Series LXXI 2, Princeton.

Bell, J. L. (1977) *Boolean Valued Models and Independence Results in Set Theory*, Oxford University Press, Oxford.

Benacerraf, P. and Putman, H. (ed.) (1964) *Philosophy of Mathematics: Selected Readings* (1st edn), Prentice-Hall, Englewood Cliffs, NJ.

Benacerraf, P. and Putnam, H. (ed.) (1983) *Philosophy of Mathematics: Selected Readings* (2nd edn), Cambridge University Press, Cambridge.

Berkeley, G. (1734) 'The analyst or a discourse addressed to an infidel mathematician.' Reprinted in *The Works of George Berkeley*, A. A. Luce and T. E. Jessop (eds) Thomas Nelson and Sons Ltd. (1948–51).

Bernoulli, Jakob (Jaques) (1713) *Ars Conjectandi*, Basle, pt. IV, ch. 5. English translation of pt. IV by Bing Sung, Harvard University Dept. of Statistics Technical Report 2 (1966).

Bishop, E. (1967) *Foundations of Constructive Analysis*, McGraw-Hill, New York.

Bolzano, B. (1851). *Paradoxien des Unendlichen*, Leipzig, tr. as *Paradoxes of the Infinite*, F. Prihonsky, Routledge and Kegan Paul, London; Yale University Press, New Haven (1950).

Boolos, G. (1971). 'The iterative conception of set.' *Journal of Philosophy*, 68; reprinted in Benacerraf & Putnam (1983).

Bostock, D. (1975) *Logic and Arithmetic Vol. 2*, Clarendon Press, Oxford.

Boyer, C. B. (1959) *The History of the Calculus and its Conceptual Development*, Dover, New York.

Brouwer, L. E. J. (1949) 'Consciousness, philosophy and mathematics.' *Proceedings of the Tenth International Congress of Philosophy (Amsterdam, 1948)* vol. 1., Pt. 2, E. W. Beth, H. J. Pos and J. H. A. Hollack (eds), Amsterdam (1952). Reprinted in Benacerraf and Putnam (1964, 1983).

Cantor, G. (1883) 'Fondaments d'une théorie générale des ensembles' *Acta Mathematica 2*, 381–408.

—— (1915/1955) *Contributions to the Founding of the Theory of Transfinite Numbers*, tr. P. E. B. Jourdain, Open Court, LaSalle, Illinois/ Dover, New York.

Carruccio, E. (1964) *Mathematics and Logic in History and Contemporary Thought*, Faber and Faber, London.

Chihara, C. (1973) *Ontology and the Vicious Circle Principle*, Cornell University Press, Ithaca, NY.

Cohen, P. J. (1966) *Set Theory and the Continuum Hypothesis*, W. A. Benjamin, New York.

Dantzig, T. (1954) *Number: The Language of Science*, Allen and Unwin, London.

Dauben. J. W. (1979) *Georg Cantor: His Mathematics and Philosophy of the Infinite*, Harvard University Press, Cambridge, Massachusetts.

Dedekind, R. (1901/1963) *Essays on the Theory of Numbers*. Tr. W. W. Beman, Open Court, Chicago/Dover, New York.

Drake, F. R. (1974) *Set Theory: An Introduction to Large Cardinals*, North Holland, Amsterdam.

Dummett, M. (1959) 'Wittgenstein's philosophy of mathematics.' *Philosophical Review*, LXVIII, 324–48. Reprinted in P. Benacerraf and H. Putnam (1964).

—— (1963) 'Realism' in Dummett (1978).

—— (1967) entry on Frege in P. Edwards (ed.) *The Encyclopedia of Philosophy*, Free Press, MacMillan, New York. Reprinted as 'Frege's philosophy' in Dummett (1978).

—— (1978) *Truth and Other Enigmas*, Duckworth, London.

Einstein, A. (1920) *The Theory of Relativity*, tr. R. W. Lawson, Methuen, London.

Feyerabend, P. (1975) *Against Method*, NLB, London.

Fitting, M. (1969) *Intuitionistic Logic, Model Theory and Forcing*, North Holland, Amsterdam.

Fraassen, B. C. van (1970) *An Introduction to the Philosophy of Space and Time*, Random House, New York.

Frege, G. (1959) *The Foundations of Arithmetic*, tr. J. L. Austin, Basil Blackwell, Oxford.

Gale, R. (1967) *The Philosophy of Time*, Anchor Books, New York.

Geach, P. T. (1962) *Reference and Generality*, Cornell University Press, Ithaca, New York.

Gödel, K. (1938) 'The consistency of the axiom of choice and the generalized continuum hypothesis.' *Proc. Nat. Acad. Sci. USA*, 24, 556–7.

—— (1939) 'Consistency proof for the generalized continuum hypothesis'. *Proc. Nat. Acad. Sci. USA*, 25, 220–4.

—— (1944/1964) 'Russell's mathematical logic.' In P. A. Schilpp (ed.) *The Philosophy of Bertrand Russell*, The Library of Living Philosophers. Tudor Pub. Co., New York. Reprinted in P. Benacerraf and H. Putnam (1964, 1983).

—— (1947) 'What is Cantor's continuum problem?' *Am. Math. Monthly*, 54, 515–25. Reprinted in P. Benacerraf and H. Putnam (1964, 1983).

Goodman, N. (1964) 'A world of individuals.' In P. Benacerraf and H. Putnam (1964).

Grant, E. (1969) 'Medieval and seventeenth-century conceptions of the extracosmic void' *Isis*, 60, p. 41–60.

Hacking, I. (1965) *The Logic of Statistical Inference*, Cambridge University Press, Cambridge.

—— (1975) *The Emergence of Probability*, Cambridge University Press, Cambridge.

Hallett, M. F. (1979) 'Towards a theory of mathematical research programmes I and II' *British Journal for the Philosophy of Science*, 30, pp. 1–25, 135–59.

—— (1984) *Cantorian Set Theory and the Limitation of Size*, Clarendon Press, Oxford.

Hatcher, W. S. (1968) *Foundations of Mathematics*, W. B. Saunders Co., Philadelphia.

Hausdorff, F. (1957) *Set Theory*, tr. J. R. Aumann *et al.*, Chelsea Publishing Co., New York.

Heijenhoort, J. van (ed.) (1967) *From Frege to Gödel*, Harvard University Press, Cambridge, Massachusetts.

Hilbert, D. (1925) 'On the infinite.' In J. van Heijenhoort (1967).

Hobbes, T. (1665) *De Corpore*, pt. I, ch. 2. English translation in I. C. Hungerland and G. R. Vick (eds) *Thomas Hobbes Part I of De Corpore* tr. A. Martinich, Abaris Books, New York (1981).

Hofstadter, D. R. (1979) *Gödel, Escher, Bach: an eternal golden braid*, Basic Books, New York.

Jech, T. (1978) *Set Theory*, Academic Press, New York.

Kleene, S. C. (1967) *Introduction to Metamathematics*, North Holland, Amsterdam.

Kneale, W. (1949) *Probability and Induction*, Clarendon Press, Oxford.

Kneale, W. and Kneale M. (1962) *The Development of Logic*, Clarendon Press, Oxford.

Kuhn, T. (1962) *The Structure of Scientific Revolutions*, University of Chicago Press, Chicago.

Lakatos, I. (1963) 'Proofs and refutations.' *British Journal for Philosophy of Science*, 1–25, 120–39, 221–43, 296–342.

Leibniz, G. (1702) 'Letter to Varignon, with a note on the "Justification of the Infinitesimal Calculus by that of Ordinary Algebra"' pp. 542–6. In *Leibniz Philosophical Papers and Letters* tr. L. E. Loemker (ed.) Reidel, Dordrecht (1969).

—— (1966) *Logical Papers*, tr. G. H. R. Parkinson (ed.) Clarendon Press, Oxford.

Levy, A. and Solovay, R. M. (1967) 'Measurable cardinals and the continuum hypothesis.' *Israel Journal of Mathematics*, 5, pp. 234–48.

Lucas J. R. (1973) *A Treatise on Time and Space*, Methuen, London.

Long, A. A. and Sedley, D. N. (1987) *The Hellenistic Philosophers*, Vol. 1, Cambridge University Press, Cambridge.

Lucas, J. R. (1973) *A Treatise on Time and Space*, Methuen, London.

Lusin, N. (1917) 'Sur la classification de M. Baire.' *Comptes Rendus de l'Académie des Sciences de Paris*, 164, pp. 91–4.

—— (1925) 'Sur les ensembles projectifs de M. Henri Lebesgue' *Comptes Rendus de l'Academie des Sciences de Paris*, 180, pp. 1572–4.

Maddy, P. (1984) 'New directions in the philosophy of mathematics.' *Philosophy of Science Association*, 2, 427–48.

Mannheim, J. H. (1964) *The Genesis of Point Set Topology*, Pergamon, London.

Mansfield, R. and Weitkamp, G. (1985) *Recursive Aspects of Descriptive Set Theory*, Oxford University Press, Oxford.

McGuire J. E. (1983) 'Space, geometrical objects and infinity: Newton and Descartes on extension.' In W. R. Shea (ed.) (1983).

Mellor, D. H. (1971) *The Matter of Chance*, Cambridge University Press, Cambridge.

Mill, J. S. (1843) *A System of Logic*, Longmans, London.

Moore, G. H. (1982) *Zermelo's Axiom of Choice: its Origins and Development*, Springer-Verlag, New York/Heidelberg/Berlin.

Nerlich, G. (1976) *The Shape of Space*, Cambridge University Press, Cambridge.

Newton I. (1934) *Principia*. A. Motte's 1729 translation of 3rd (1726) edition revised by F. Cajori, University of California Press, Berkeley and Los Angeles.

North, J. D. (1983) 'Finite and otherwise. Aristotle and some seventeenth-century views.' In W. R. Shea (ed.) (1983).

Ockham, W. (1974) *Ockham's Theory of Terms*, Part I of the *Summa Logicae* tr. and introduction M. J. Luce, University of Notre Dame Press, Notre Dame, Indiana.

Owen, G. E. L. (1957) 'Zeno and the mathematicians.' In *Proceedings of the Aristotelian Society*, 1957–8.

Parsons, C. (1974) 'Sets and classes'. *Nous*, Vol. 8.

—— (1977) 'What is the iterative conception of set?' *Proceedings of the Fifth International Congress of Logic Methodology and Philosophy of Science, 1975, Pt. I: Logic, Foundations of Mathematics and Computability Theory*, R. E. Butts and J. Hintikka (eds), Reidel, Amsterdam. Reprinted in Benacerraf and Putnam (1983).

Pascal, B. (1954) *Oeuvres Complètes*, J. Chevalier (ed.), Bibliothèque de la Pléiade, NRF.

Putnam, H. (1967) 'The thesis that mathematics is logic.' In R. Schoenman (ed.) *Bertrand Russell: Philosopher of the Century*, Allen and Unwin, London. Reprinted in Putnam (1975).

—— (1975) *Philosophical Papers vol. I. Mathematics, Matter and Method*, Cambridge University Press, Cambridge.

Quine, W. V. O. (1953) 'Two dogmas of empiricism.' In *From a Logical Point of View*, Harvard University Press, Cambridge, Massachusetts.

—— (1966) *The Ways of Paradox and Other Essays*, Random House, New York.

—— (1963) *Set Theory and its Logic*, Harvard University Press, Cambridge, Massachusetts.

Ramsey, F. P. (1926) 'The foundations of mathematics.' *Proceedings of the London Mathematical Society 1925*, 2nd series, vol. 25, pt. 5. Reprinted in R. B. Braithwaite (ed.) *The Foundations of Mathematics and Other Essays*, Routledge and Kegan Paul, London, Harcourt Brace, New York (1931).

Rotman, B. and Kneebone, G. T. (1966) *The Theory of Sets and Transfinite Numbers*, Oldbourne, London.

Rucker, R. v. B. (1982) *Infinity and the Mind: The Science and Philosophy of the Infinite*, Birkhäuser, Boston.

Russell, B. (1903) *Principles of Mathematics*, Allen and Unwin, London.

—— (1907) 'On some difficulties in the theory of transfinite numbers and order types.' *Proceedings of the London Mathematical Society*, 2nd series, 4, 29–53.

—— (1908) 'Mathematical logic as based on the theory of types'. *American Journal of Mathematics*, 30, 222–62. Reprinted in J. van Heijenhoort (1967).

—— (1919) *Introduction to Mathematical Philosophy*, Allen and Unwin, London.

Salmon, W. C. (1967) *Zeno's Paradoxes*, Bobbs-Merrill, New York.

Shea, W. R. (ed.) (1983) *Nature Mathematized*, Reidel, Dordrecht, Holland.

Shepherdson, J. C. (1951–3) 'Inner models for set theory' Pt. I. *Journal of Symbolic Logic*, 16, 161–90; Pt. II, 17, 225–37; Pt. III, 18, 145–67.

Souslin, M. (1917) 'Sur une definition des ensembles mesurable *B* sans nombres transfinis.' *Comptes Rendus de l'Académie des Sciences de Paris*, 164, 88–91.

Tamas, G. (1986) *The Logic of Categories: Boston Studies in the Philosophy of Science*, Vol. 85, Reidel, Dordrecht.

Tragesser, R. (1984) *Husserl and Realism in Logic and Mathematics*, Cambridge University Press, Cambridge.

Troelstra, A. S. (1969) *Principles of Intuitionism*, Springer, Berlin.

—— (1977) *Choice Sequences: a Chapter of Intuitionist Mathematics*, Clarendon Press, Oxford.

Vilkenkin, N. Y. (1968) *Stories about Sets*, tr. Scripta Technica, Academic Press, New York/London.

Whitehead, A. N. and Russell, B. (1910–13) *Principia Mathematica*, 3 vols., Cambridge University Press, Cambridge.

Wittgenstein, L. (1914) 'Notes dictated to G. E. Moore in Norway.' Appendix II of *Ludwig Wittgenstein: Notebooks 1914–16*, 2nd edn, Basil Blackwell, Oxford (1979).

—— (1922) *Tractatus Logico-Philosophicus*, Routledge and Kegan Paul, London.

—— (1967) *Remarks on the Foundations of Mathematics*, Basil Blackwell, Oxford.

—— (1974) *Philosophical Grammar*, Basil Blackwell, Oxford.

Wright, C. (1980) *Wittgenstein on the Foundations of Mathematics*, Duckworth, London.

—— (1983) *Frege's Conception of Numbers as Objects*, Aberdeen University Press, Aberdeen.

Zermelo, E. (1904) 'Proof that every set can be well-ordered.' In van Heijenhoort (1967), 139–41.

—— (1908) 'A new proof of the possibility of a well-ordering.' In Heijenhoort (1967), pp. 183–198.

Glossary of Symbols

¬	sentence negation
∧	and
∨	or
⇒	if ... then
⇔	if and only if
≡	equivalence
$\exists x$	there is an x such that ...
$\forall x$	for every x ...
<	less than
$A \vdash B$	B is deducible from A
iff	if and only if
$x \in y$	x is an element of y
$x \notin y$	x is not an element of y
$x \subseteq y$	x is included in y
∅	null set
$\{a\}$	set whose only member is a
$\{a, b\}$	set whose members are a and b
$\langle x, y \rangle$	ordered pair of x and y
$\{x : Fx\}$	class of objects x such that Fx
$A \cup B$	union of sets A and B
A^B	set of functions from B to A
$\bigcup_{\beta < \alpha} A_\beta$	union of sets A_β for $\beta < \alpha$
$P(A)$	power set of A
$C(A)$	the power or cardinality of A
$O(A, \leqslant)$	the order type of A ordered by $\leqslant$
ω	first infinite ordinal number
ω_1	second infinite ordinal number

$\aleph_0$	aleph zero, first infinite cardinal number
$\aleph_1$	cardinal number of second number class $(=\omega_1)$
$2^{\aleph_0}$	cardinal number of set of real numbers
$On(\alpha)$	α is an ordinal number
$\alpha, \beta, \ldots$	variables used for ordinal numbers
$\phi, \psi, \ldots$	variables used for propositional functions
$\phi!$	predicative propositional function
V_α	set of all set of rank α (see p. 157)
L_α	set of all constructible sets of rank α (see p. 177)
M_α	set of all strongly constructible sets of rank α (see p. 184)
AC	axiom of choice (p. 123)
AD	axiom of determinacy (p. 205)
DD	axiom of definable determinacy (p. 206)
CH	Continuum Hypothesis – $2^{\aleph_0} = \aleph_1$
GCH	Generalised Continuum Hypothesis – $2^{\aleph_\alpha} = \aleph_{\alpha+1}$
V = L	all sets are constructible
ZF	Zermelo Fraenkel axioms for set theory

Index